# MINERALS

Edited by **Khalid S. Essa**

**Minerals**
http://dx.doi.org/10.5772/intechopen.74902
Edited by Khalid S. Essa

**Contributors**

Aref Alshameri, Fuguo Yang, Wei Xinghu, Xin Chen, Wang Hailong, Yan Chunjie, Xu Feng, He Hongping, Basanta Sahu, Arkoprovo Biswas, Francisco Colmenero, Khalid S. S. Essa, Marc Munschy



First published in London, United Kingdom, 2019 by IntechOpen
IntechOpen is the global imprint of INTECHOPEN LIMITED, registered in England and Wales, registration number: 11086078, 7th floor, 10 Lower Thames Street, London,
EC3R 6AF, United Kingdom
Printed in Croatia

British Library Cataloguing-in-Publication Data
A catalogue record for this book is available from the British Library

Additional hard and PDF copies can be obtained from orders@intechopen.com

Minerals, Edited by Khalid S. Essa
p. cm.
Print ISBN 978-1-83962-681-4
Online ISBN 978-1-83962-682-1
eBook (PDF) ISBN 978-1-83962-683-8

For EU product safety concerns:
IN TECH d.o.o., Prolaz Marije Krucifikse Kozulić 3, 51000 Rijeka, Croatia,
info@intechopen.com or visit our website at intechopen.com.

# Meet the editor

Dr. Essa received his B.Sc. degree with honors (1997), an M.Sc. degree (2001) and a Ph.D. degree (2004) in Geophysics from the Faculty of Science, Cairo University. He joined the staff of Cairo University (1997) and was appointed a Research Professor of potential field methods in the Department of Geophysics (2014). He has undertaken post-doctoral visits to Strasbourg University, France (2018), Charles University in Prague, Czech (2014), and Western Michigan University, USA (2006) and served as an editor and external reviewer for many top journals. He has attended several international geophysical conferences in USA, Australia, and France. He is a member of SEG, AGU, AAPG, EAGE, and EGS. He is also a member of the Petroleum and Mineral Resources Research Council, Academy of Scientific Research and Technology, Egypt (2018). He was awarded the Award of the Prof. Nasry Matari Shokry in Applied Geology, Academy of Scientific Research & Technology (2017) and the Award of Cairo University for Scientific Excellence in Interdisciplinary, Multidisciplinary and Future Sciences (2017).

# Contents

# Preface

Mineral exploration (and the bearing rocks) is a catalyst for increased investment and income in the world. Using many different geophysical and geological methods, it is easy to find minerals in different places around the world. The continuous development of different methods for finding and handling mineral-bearing rocks using different physical and chemical approaches has led to a huge revolution in the knowledge of various micro-mineral components, which has increased the importance of minerals.

This book consists of various chapters written by scientists and researchers from different countries. This book is divided into five chapters organized into two basic parts: (1) dealing with mineral exploration via geophysical methods and (2) the importance of some economic mineral deposits. The book includes an introductory chapter on mineral exploration from the point of view of geophysicists. The first part of this book includes two chapters to illustrate the importance of using geophysical methods to investigate minerals and their bearing rocks. The second part contains three chapters including the importance of clay and the several minerals contained within (with a very small particle size), as well as uranium, which contains radioactive elements and has an importance in safe nuclear energy and in generating electricity in different countries. Finally, the law of distribution and the formation of minerals and ore in rocks from the mining stage and extraction to the marketing process.

This book will hopefully inspire readers, researchers, geologists, geophysicists, and the mining community to look further into the frontier topics of minerals and it opens new possible research paths for further development.

I would like to thank Almighty Allah for helping me with this book. It is my most pleasant duty to express my sincere gratitude to my parents' souls, brothers and sisters whose contribution cannot be put in words but can only be felt deeply in the heart. I would also like to thank my wife (Prof. Amany Fekry), my daughters (Sama, Salma, Sally), my colleagues, and all those who have credited me for completing this work in the appropriate style. I would also like to thank Prof. Dr. Mohamed Othman El Khosht, Cairo University President and Prof. Dr. Abd El Hamid Wagdi El Manawi, Dean of the Faculty of Science, for their continuous support in our scientific and practical life. Finally, I wish to express my sincere gratitude to the publisher and Mrs. Jasna Božić, Author Service Manager, for her patience during this project.

**Khalid S. Essa**
Faculty of Science
Cairo University, Egypt

Section 1

# Mineral Exploration via Geophysical Methods

Chapter 1

# Introductory Chapter: Mineral Exploration from the Point of View of Geophysicists

Khalid S. Essa and Marc Munschy

Additional information is available at the end of the chapter

http://dx.doi.org/10.5772/intechopen.84830

## 1. Introduction

Mineral exploration is vital in many countries to increase the income of their people, and their economy relies upon discovering minerals. The minerals excavated are iron, copper, gold, silver, molybdenum, zinc, coal, uranium, sulfide, tin, chromite, potash, etc. From the point of view of geophysicists, geophysical methods are playing an important role in mineral investigation, groundwater investigation and hydrocarbon exploration [1–3]. Geophysical methods are grouped into two different kinds which are passive and active methods. Passive techniques measured the Earth's natural fields as gravity, magnetic and self-potential (SP), while active methods distinguish variabilities of physical parameters in the Earth's layers produced by non-natural sources like seismic, electrical resistivity, induced polarization methods, etc. Various geophysical techniques rely upon different physical properties in the subsurface or deeper. The selection of a particular method relies on various parameters including cost, efficiency, accessibility, and type of application. In addition, a single choice of a geophysical method in any application occasionally provides poorly constrained results. So, a combination of two or more approaches certifies much more consistent results. This methodology is called integrated geophysical approach that ensures more prominent precision and higher consistency of results. It has to be emphasized that geophysical models are generally not unique regarding geometry (shape, size, and depth) of the buried structures. The spatial location of the buried sources and their depth can also be precisely assessed by some mathematical ways [4].

## 2. Geophysical methods: selection and objectives

The worth of geophysical techniques in mineral investigation relies on the variability of physical properties as well as on local geological environment, topography, etc. Each region

| Geophysical methods | Measured field | Physical properties | Units | Typical minerals and applications | Exploration depth |
|---|---|---|---|---|---|
| Gravity | Natural gravity field of the Earth | Density | mGal | Sulfides including sphalerite, barite, mining, hydrology, plate tectonics | All |
| Magnetic | Natural magnetic field of the Earth | Magnetic susceptibility<br>Remanent magnetization | nT | Magnetite, ultramafics, iron-rich rocks, basin analysis, plate tectonics | Until curie isotherm |
| Self-potential | Natural telluric current | Electrical conductivity | mV | Metallic sulfides, serpentinite, graphite, water-filled shears, salt water | A few hundred meters |

**Table 1.** Main natural geophysical methods and their essentials in mineral exploration.

tends to have its own distinct geophysical identity requiring the adoption of appropriate geophysical methods. Typically, more than one method is used in one survey, to decrease the ambiguity in deciphering the nature of unknown resources below the surface. Some important geophysical methods (**Table 1**) have been chosen in this chapter to demonstrate how we use these methods in mineral exploration.

## 3. Gravity method

The gravity method studies anomalies of the Earth's gravitational field due to changes in densities below the surface. Density changes (density contrast) are induced by an occurrence of

| Mineral/rock | Density (g $cm^{-3}$) | Magnetic susceptibility ($10^{-6}$ SI) |
|---|---|---|
| Gold | 19.28 | −0.14 |
| Chromite | 4.80 | 3000–120,000 |
| Iron | 7.87 | 3,900,000 |
| Copper | 8.90 | −9.63 |
| Graphite | 2.16 | −80 to 200 |
| Hematite | 5.26 | 500–40,000 |
| Magnetite | 5.20 | 1,000,000–5,700,000 |
| Pyrite | 5.01 | 35–5000 |
| Sphalerite | 4.08 | −31 to 750 |
| Rock salt | 2.5–2.6 | $-0.01 \times 10^{-3}$ |
| Dolomite | 2.87 | −10 to 940 |
| Granite | 2.64 | 0–50,000 |

**Table 2.** Density and magnetic susceptibility of some minerals and rocks.

a causative body (target source) within the surrounding rocks. Rock densities are considered as one of the variables of all geophysical parameters. The density of rocks is dependent on both mineral composition and porosity. **Table 2** shows examples of some mineral and rock density values. Gravity method can be used from the land to the air and in marine environment. Gravity anomalies are due to anomalous density within the Earth. Gravity method has wide-ranging uses in mineral, hydrocarbon, cave, geothermal, and archeological investigations [5–7]. The target of gravity interpretation is to locate and characterize the buried mineral source parameters, in particular, the density contrast, depth, and shape [8–10]. Several methodologies are used to interpret gravity data [11–14]. More recently, three-dimensional modeling and inversion of gravity data provide more accurate results. In these days, new nonconventional methods are used such as particle swarm optimization, very fast simulated annealing, genetic algorithm, forced neural network, and differential evolution algorithm. **Table 2** demonstrates value of the density of rock and mineral examples.

## 4. Magnetic method

Magnetic method is one of the oldest branches of geophysics and used in many exploration issues such as mineral and ores as massive sulfide, iron, gold, and porphyry copper deposits. Magnetic data interpretation has shown its efficiency in the identification of deep and shallow structures known to employ a structural control on mineralization occurrences [15, 16]. Uniform geological models (geologic contacts, thin sheets, cylinders, and spheres) are frequently employed in magnetic inversion to estimate the body factors (the amplitude factor, the depth, the index angle, the location of the origin, and the shape) and have a vital role in many exploration issues. These models cannot be an exact geologically representation, but are generally a good tool in magnetic interpretation to calculate, in particular, the body parameters. Several elucidation approaches of the magnetic data above inhomogeneous geological structures have been recognized. These approaches can be characterized into four categories as follows: Category I is the well-known two- and three-dimensional magnetic modeling and inversion for irregular structures. Category II is recognized by using residual magnetic anomalies only. Category III is relied upon using not only the residual but also the measured magnetic data. Category IV is dependent on utilizing the metaheuristic algorithms like the particle swarm optimization (PSO) method, the genetic algorithm (GA) method, the differential evolution algorithm (DEA) method, the simulated annealing algorithm (SAA) method, the ant colony optimization (ACO) method, and the neural network (NN) method. Magnetization directionally consists of adding induced and remanent components. Induced magnetization depends on the magnetic susceptibility of the material (**Table 2**) and the magnitude and direction of the Earth's magnetic field, whereas remanent magnetization reflects the past magnetic history of the material.

## 5. Self-potential method

Self-potential (SP) is one of the passive geophysical techniques that measure the natural Earth's surface electric potential happening by many reasons like the difference between minerals and their hosting, bioelectric generation from plants, and electrochemical and electrokinetic. Sato and Mooney [17] demonstrated that this potential has different types as electrokinetic potential produced from the stream of a fluid with certain electrical properties going through a pipe or permeable medium with various electrical properties. In addition, several another mechanisms (diffusion, mineralization, etc.) produced this SP. The self-potential method has a wide range in different applications in exploration of geophysics, hydrogeophysics, and environmental problems and mineral exploration such as metallic sulfides, magnetite, graphite, and uranium. Several assessable elucidation approaches of the SP data over the buried geologic structures have been established. These approaches can be classified into two categories. The first category is usually dependent on using simple geological models (spheres and cylinders) to appraise the parameters for buried structures and has a vigorous role in many investigation problems as linear and nonlinear least squares methods, moving average and gradient methods, depth-horizontal curve method. These models are not wholly geologically perfect, but they are often useful in SP interpretation to calculate the body parameters. The second category is dependent on two- and three-dimensional modeling and inversion methods. However, a portion of these methods requires good initial parameters, using a few data point and distances, and requires more time.

Finally, the three potential methods (gravity, magnetic, and self-potential) mentioned above have been used to evaluate the source parameters but are suffering of ill-posedness and nonuniqueness in finding a global solution [18]. The usage of simple geometrical structures in gravity, magnetic, and self-potential inversion helps in overcoming some of these limitations, gives an optimal fit for the buried structures, and plays a vigorous role in solving many investigation problems.

## 6. Case studies

### 6.1. Gravity anomaly of chromite deposit body

**Figure 1** shows the gravity anomaly of length 180 m over a chromite deposit body in the chromite region of the Camaguey Area, Cuba [19]. This chromite deposits are found in a complex geological environment involving serpentinized peridotite and dunite with slight quantities of gabbro, troctolite, and anorthosite. This complex environment affected by metamorphic rocks and superimposed by upper Cretaceous volcanic rocks with limestone and radiolarian cherts. Severe compressive stresses, started in late Cretaceous or early Eocene time, deformed both the sedimentary rocks and the underlying ultramafic complex and culminated in extensive thrust faulting, probably in the late middle Eocene. Uplift and erosion have detached the overlying rocks from the serpentine except in synclinal areas, the largest of which extends from Central Lugareño to Loma Yucatan. This gravity anomaly has been interpreted by utilizing different inversion methods as demonstrated in **Table 3**. **Table 3** demonstrated that the estimated chromite deposit body parameters, amplitude factor (A), depth (z), location of the body (d), and the shape (q) by utilizing these approaches, have a reasonable agreement especially the depth with that obtained from drilling.

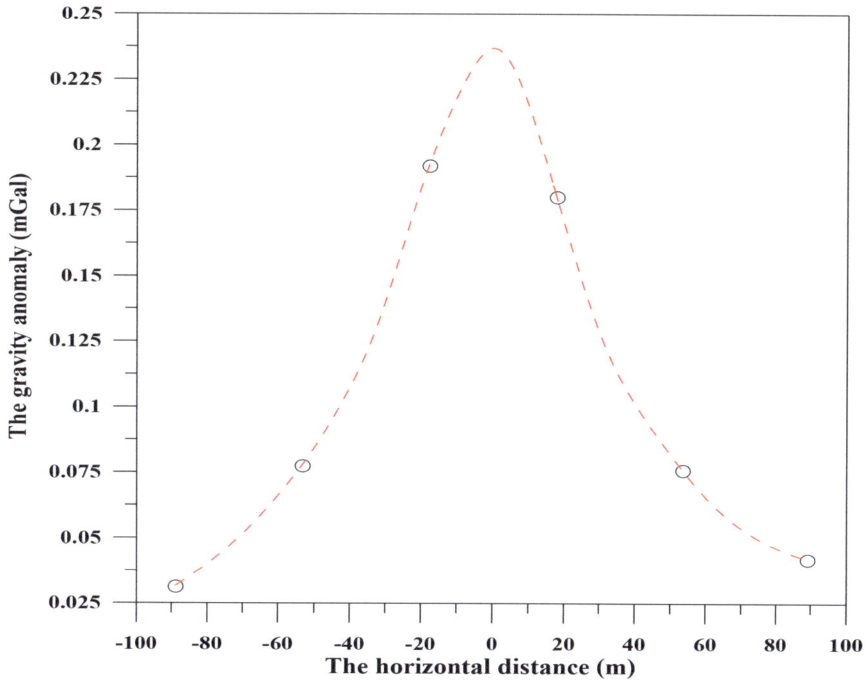


**Figure 1.** The gravity anomaly of chromite deposit body (Cuba).

### 6.2. Magnetic anomaly over an olivine diabase dike

**Figure 2** demonstrates the magnetic anomaly profile of length 2200 m above an olivine diabase dike from the Pishabo Lake, Canada, and this site is made out of plagioclase, purplish-brown augite, pale green olivine, apatite, some biotite, and large patches of magnetite [20]. This magnetic anomaly has been interpreted by using various inversion algorithms such as moving average method, parametric inversion method, and the PSO method. The elucidation procedure and their produced results are mentioned in **Table 4**. The predicted parameters (M which represents the amplitude factor, z is the depth, θ is the magnetization angle, d is the origin location, and q is the shape) of the body by using these inversion methods have a good agreement together.

| Parameters | Drilling information | Essa method [21] | Biswas method [9] | Ekinci et al. method [22] | Essa and Munschy method [23] |
|---|---|---|---|---|---|
| A (mGal m$^2$) | — | 412.33 | 16.80 | 288.25 | 408.25 |
| z (m) | 21.00 | 21.02 | 42.30 | 23.23 | 21.15 |
| d (m) | — | — | −2.40 | 58.73 | 0.63 |
| q (dimensionless) | — | 1.5 | 1.0 | 1.5 | 1.47 |

**Table 3.** The results obtained for interpreting gravity anomaly of chromite deposit body (Cuba) using different inversion methods.

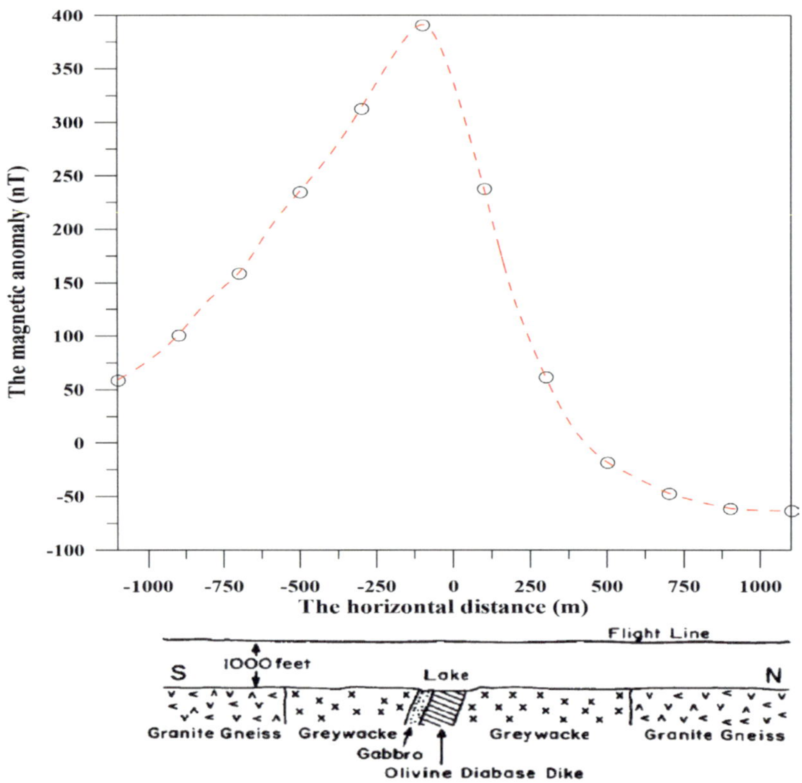


**Figure 2.** The magnetic anomaly over an olivine diabase dike from the Pishabo Lake, Canada.

### 6.3. Self-potential anomaly of sulfide orebody

**Figure 3** displays the self-potential anomaly of a sulfide orebody in the Sariyer area which is located about 18 km north to Istanbul, Turkey, and characterized by an outcropping of andesite, pyrite veins, and cupriferous waters. The area of this investigation is characterized by a steep surface gradient. In 1951, the sulfide orebodies had been explored by utilizing

| Parameters | McGrath and Hood method [20] | Abdelrahman et al. method [24] | Abdelrahman et al. method [25] | Biswas method [26] | Essa method |
|---|---|---|---|---|---|
| M (nT) | — | — | 1429.0 | 1411.8 | 1380.8 |
| z (m) | 304 | 318.9 | 320 | 324 | 305.9 |
| θ (°) | — | — | 37.5 | −37.9 | 38.5 |
| d (m) | — | 2.86 | — | 1.7 | 2.22 |
| q (dimensionless) | — | 1.0 | 1.0 | 1.0 | 0.98 |

**Table 4.** The results obtained for interpreting magnetic anomaly over an olivine diabase dike from the Pishabo Lake, Canada.

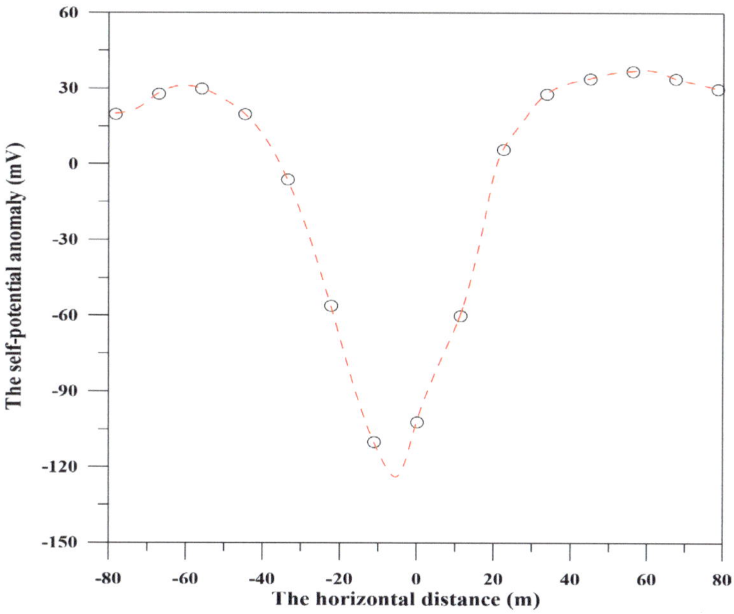


**Figure 3.** The self-potential anomaly of a sulfide orebody in the Sariyer Area, Turkey.

| Parameters | Abdelrahman et al. method [27] | Asfahani and Tlas method [28] | Biswas and Sharma method [29] | Essa method |
|---|---|---|---|---|
| K (mV × m²) | 3245.0 | 4695.6 | 2855.4 | 3447.1 |
| z (m) | 24.5 | 22.6 | 28.0 | 25.4 |
| θ (°) | −51.2 | −82.9 | −70.3 | −58.9 |
| d (m) | — | — | −2.5 | −0.5 |
| q (dimensionless) | 1.5 | 1.0 | 1.5 | 1.5 |

**Table 5.** The results obtained for interpreting self-potential anomaly of a sulfide orebody in the Sariyer Area, Turkey.

geophysical techniques where it found to lie under unmineralized schist or alluvium with a depth of 23 m and elongated as a spheroid dimension [21]. This profile has a length of 160 m and has been subjected to many interpretation methods to estimate the sulfide orebody parameters (K is the amplitude factor, z is the depth, θ is the polarization angle, d is the origin location, and q is the shape). The estimated results are displayed in **Table 5**. The estimated parameters of this source by exploiting these methods have a good covenant together.

## 7. Conclusions

The chapter discussed the importance of the geophysical methods, especially gravity, magnetic, and self-potential methods, in mineral and ore exploration which are considered as an important issue for many countries to increase their incomes. The results of the published information described in the state of arts mentioned above by the three case studies revealed the pervasiveness of these methods and its capability of elucidating gravity, magnetic, and self-potential data associated with shallow and deep mineralized bodies.

## Acknowledgements

The authors would like to thank and express appreciation to Ms. Jasna Bozic, Author Service Manager, for her assistance and cooperation in this issue. The first author wishes to thank the Science and Technology Development Fund (STDF) and the Institut Francais d'Egypte (IFE) for providing their full support to finish this work.

## Conflict of interest

There is no conflict of interest.

## Author details

Khalid S. Essa[1]* and Marc Munschy[2]

*Address all correspondence to: khalid_sa_essa@yahoo.com

1 Geophysics Department, Faculty of Science, Cairo University, Giza, Egypt

2 Institut de Physique du Globe de Strasbourg, EOST, CNRS, University of Strasbourg, Strasbourg, France

## References

[1] Haldar S, editor. Mineral exploration: Principles and Applications. 1st ed. Netherlands: Elsevier; 2018. 378 p

[2] Abdelrahman EM, El-Araby TM, Essa KS. A least-squares minimisation approach to depth, index parameter, and amplitude coefficient determination from magnetic anomalies due to thin dykes. Exploration Geophysics. 2003;**34**:241-248

[3] Khalil MH, Ahmed KS, Elnahry AH, Hasan AN. Integrated geophysical, remote sensing and GIS studies for groundwater assessment, Abu Zenima Area, West Sinai, Egypt. International Journal of Geosciences. 2014;**5**:882-907

[4] Chen Z, Mou L, Meng X. The horizontal boundary and top depth estimates of buried source using gravity data and their applications. Journal of Applied Geophysics. 2016; **124**:62-72

[5] Essa KS. A fast interpretation method for inverse modeling of residual gravity anomalies caused by simple geometry. Journal of Geological Research. 2012;**2012**:327037

[6] Hinze WJ, von Frese RRB, Saad AH. Gravity and Magnetic Exploration—Principles, Practices, and Applications. 1st ed. Cambridge University Press; 2013. 525 p

[7] Deng Y, Chen Y, Wang P, Essa KS, Xub T, Liang X, et al. Magmatic underplating beneath the Emeishan large igneous province (South China) revealed by the COMGRA-ELIP experiment. Tectonophysics. 2016;**672-673**:16-23

[8] Essa KS. New fast least-squares algorithm for estimating the best-fitting parameters of some geometric-structures to measured gravity anomalies. Journal of Advanced Research. 2014;**5**:57-65

[9] Biswas A. Interpretation of residual gravity anomaly caused by a simple shaped body using very fast simulated annealing global optimization. Geoscience Frontiers. 2015;**6**: 875-893

[10] Mehanee SA, Essa KS. 2.5D regularized inversion for the interpretation of residual gravity data by a dipping thin sheet: Numerical examples and case studies with an insight on sensitivity and non-uniqueness. Earth, Planets and Space. 2015;**67**:130

[11] Kilty KT. Werner deconvolution of profile potential field data. Geophysics. 1983;**48**:234-237

[12] Stavrev PY. Euler deconvolution using differential similarity transformations of gravity or magnetic anomalies. Geophysical Prospecting. 1997;**45**:207-246

[13] Abdelrahman EM, Abo-Ezz ER, Essa KS, El-Araby TM, Soliman KS. A least-squares variance analysis method for shape and depth estimation from gravity data. Journal of Geophysics and Engineering. 2006;**3**:143-153

[14] Abdelrahman EM, Essa KS. A new approach to semi-infinite thin slab depth determination from second moving average residual gravity anomalies. Exploration Geophysics. 2013;**44**:185-191

[15] Essa KS, Nady AG, Mostafa MS, Elhussein M. Implementation of potential field data to depict the structural lineaments of the Sinai Peninsula, Egypt. Journal of African Earth Sciences. 2018;**147**:43-53

[16] Le Maire P, Munschy M. 2D potential theory using complex algebra: New equations and visualization for the interpretation of potential field data. Geophysics. 2018;**83**:J1-J13

[17] Sato M, Mooney HM. The electrochemical mechanism of sulfide self-potentials. Geophysics. 1960;**25**:226-249

[18] Tarantola A. Inverse Problem Theory and Methods for Model Parameter Estimation. 1st ed. Philadelphia: Society for Industrial and Applied Mathematics (SIAM); 2004. 352 p

[19] Roy L. Short note: Source geometry identification by simultaneous use of structural index and shape factor. Geophysical Prospecting. 2001;**49**:159-164

[20] McGrath PH, Hood PJ. The dipping dike case: A computer curve-matching method of magnetic interpretation. Geophysics. 1970;**35**:831-848

[21] Essa KS. A new algorithm for gravity or self-potential data interpretation. Journal of Geophysics and Engineering. 2011;**8**:434-446

[22] Ekinci Y, Balkaya C, Göktürkler G, Turan S. Model parameter estimations from residual gravity anomalies due to simple-shaped sources using differential evolution algorithm. Journal of Applied Geophysics. 2016;**129**:133-147

[23] Essa KS, Munschy M. Gravity data interpretation using the particle swarm optimization method with application to mineral exploration. Journal of Earth System Science. 2019; In press

[24] Abdelrahman EM, Soliman KS, El-Araby TM, Abo-Ezz ER, Essa KS. A least-squares standard deviation method to interpret magnetic anomalies due to thin dikes. Near Surface Geophysics. 2009;**7**:41-46

[25] Abdelrahman EM, Abo-Ezz ER, Essa KS. Parametric inversion of residual magnetic anomalies due to simple geometric bodies. Exploration Geophysics. 2012;**43**:178-189

[26] Biswas A. Interpretation of gravity and magnetic anomaly over thin sheet-type structure using very fast simulated annealing global optimization technique. Modeling Earth Systems and Environment. 2016;**2**:30

[27] Abdelrahman EM, Saber HS, Essa KS, Fouda MA. A least-squares approach to depth determination from numerical horizontal self-potential gradients. Pure and Applied Geophysics. 2004;**161**:399-411

[28] Asfahani J, Tlas M. Interpretation of self-potential anomalies by developing an approach based on linear optimization. Geosciences and Engineering. 2016;**5**:7-21

[29] Biswas A, Sharma SP. Interpretation of self-potential anomaly over idealized body and analysis of ambiguity using very fast simulated annealing global optimization. Near Surface Geophysics. 2015;**13**:179-195

Chapter 2

# Inversion of Amplitude from the 2-D Analytic Signal of Self-Potential Anomalies

Arkoprovo Biswas

Additional information is available at the end of the chapter

http://dx.doi.org/10.5772/intechopen.79111

**Abstract**

In the present study, analytic signal amplitude (ASA) or total gradient (TG) inversion of self-potential anomalies has been carried out using very fast simulated annealing (VFSA) global optimization technique. The results of VFSA optimization demonstrate the application and efficacy of the proposed method for idealized synthetic hypothetical models and real single and multiple geological structures. The model parameters deciphered here are the amplitude coefficient ($k$), horizontal location ($x_0$), depth of the body ($z$), and shape ($q$). Inversion of the model parameter suggests that constraining the horizontal location and the shape factor offers the most reliable results. Investigation of convergence rate, histogram, and cross-plot examination suggest that the interpretation method developed for the self-potential anomalies is stable and the model parameters are within the estimated ambiguity. Inversion of synthetic noise-free and noise-corrupted data for single structures and multiple structures in addition to real field information exhibits the viability of the method. The model parameters estimated by the present technique were in good agreement with the real parameters. The method has been used to invert two field examples (Sulleymonkoy anomaly, Ergani, Turkey, Senneterre area of Quebec, Canada) with application of subsurface mineralized bodies. This technique can be very much helpful for mineral or ore bodies investigation of idealized geobodies buried within the shallow and deeper subsurface.

**Keywords:** self-potential, total gradient, idealized bodies, VFSA, uncertainty estimation, ore exploration

## 1. Introduction

The self-potential (SP) technique has an important significance in mineral and ore explorations [1–8]. The method has an extensive range of application, viz., mining industries [9–12], sulfide, graphite exploration, and groundwater exploration [13, 14], study of groundwater flow in pumping wells [15, 16], geothermal exploration [17–19], fluid flow in the vadose zone [20, 21], uranium mineralization [2–4], engineering and environmental applications [22–24], archeological investigations [25], cave detection [26], earthquake prediction [27], hydraulic fracturing [28], brine contamination [29], spring flow [30], delineation of buried paleochannels [31], volcanic eruptions [32, 33], and paleoshear zones [34].

In general, most of the interpretation of SP anomaly is carried out assuming simple geometrical shapes (for example, 3-D sphere, 2-D horizontal or vertical cylinder, and 2-D sheet) buried at variable depth generated by different mineral or ore bodies [2, 3, 7, 34–47]. Numerous interpretation techniques have been developed over the last two decades considering various interpretation methods [38, 40, 48–53]. For instance, graphical methods [42, 54], the nomograms [42], logarithmic curve matching [41, 55], characteristic points, [43, 45, 56], least square method [38, 57], Fourier analysis [58], 3-D topography effect [59, 60], gradient and derivative study [61], moving average residual anomalies [62], modular neural networks [49], particle swarm optimization [63], depth from extreme point [64], differential evolution [65], Genetic-Price algorithm [66], spectral and tomographic approach [67], second horizontal gradient [68], and spectral methods [69] were too applied for the elucidation of SP data. A detailed review of the SP background, theoretical modeling, inversion, and its application in mineral exploration can be found after Biswas [1].

Among various interpretation and inversion techniques mentioned above, for the most part, inverse modeling intends to do the best interpretation of the model parameters. It is very familiar in geophysical data that the nonuniqueness problem, nonlinearity, and ill-posed nature of the SP anomaly inversion make the processing and elucidation rather cumbersome. Also, in linear problems for SP data, the presence of less number of well-known numbers rather than the number of unidentified geological model parameters often shows an uncertainty, which creates the interpretation quite difficult [70]. Subsequently, the inversion problems of SP anomaly firmly need a few limitations, *a priori* information with a specific aim to improve interpretable and reasonable model parameters [1]. However, in almost every case, the measured SP anomaly was interpreted without considering the analytical signals or total gradient derived from SP measurements.

The main objective of the present work is to interpret the analytic signal amplitude (ASA) or the total gradient (TG) of SP anomaly over different idealized or causative body, which satisfies the Laplace's condition. It is well known that the analytical signals derived from SP anomaly are correctly known as the TG [71–74]. The interpretation of ASA or TG derived from SP has been sparsely done in the present literature. It was only interpreted using ant colony optimization (ACO) [75]. In the current work, very fast simulated annealing (VFSA) was applied to decide the different model parameters associated with idealized subsurface bodies for ASA or TG from SP anomalies. The present inversion method has an advantage over other approaches for its pliability and its proficiency to converge toward global optima. The method has an ability to avoid from getting stuck in local minima and it has very high resolution, faster calculation as well as less memory without negotiating the resolution of the model parameters [2, 3, 35, 76–80]. Moreover, it does not need a priori information for the elucidation of SP anomalies.

In this study, inversion using VFSA algorithm was performed with the help of synthetic noise-free and noise-corrupted synthetic data using single and multiple structures and two field data from Sulleymonkoy anomaly, Ergani, Turkey, Senneterre area of Quebec, Canada. The results from the present method were compared with other well-established SP anomaly interpretations, such as ACO techniques. The present method used as a comprehensive method for quantitative elucidation of SP anomalies derived after various subsurface idealized geobodies.

## 2. Methodology

### 2.1. Forward modeling

Following Nabighian [71], the 2-D analytical signal amplitude (ASA) or the total gradient (TG) of SP anomaly is given as.

$$ASA(x) = (V_x^2 + V_y^2)^{1/2} \tag{1}$$

where $V_x$ and $V_y$ are the horizontal and vertical derivatives of the SP anomalies, respectively. The 2-D ASA of few of the idealized geobodies can be estimated following the general expression of SP anomaly *ASA (x)* for idealized causative bodies (horizontal/vertical cylinder, sphere, inclined sheet, etc.) at any point on the surface of the earth (**Figure 1**) [70, 71, 74, 75], which is given by the equation:

$$ASA(x) = k\left[\frac{1}{[(x-x_0)^2+(z)^2]^q}\right] \tag{2}$$

where $k$ is the amplitude coefficient/factor related to the physical properties of the source, $z$ is the depth from the surface to the top of the body (sphere, cylinder, sheet, line of poles, point pole), $x_0$ (i = 1, …,N) is the horizontal position coordinate on the surface, and $q$ is the shape factor. The shape factors ($q$) for horizontal/vertical cylinder, sphere, inclined sheet, line of poles, and point poles are 1, 1.5, 1, 0.5, and 1, respectively. The total derivation of the ASA or TG anomaly can be found in various literatures [71–75, 81].

The above equation can be used to interpret single structure. However, in order to interpret the numerous structures, the expression (2) can be rewritten as [36]:

$$ASA(x_i) = \sum_{j=1}^{M} ASA_j(x_i) \tag{3}$$

where $ASA_j(x_i)$ is the SP anomaly at $x_i$ location for $j$th body and $M$ is the number of bodies.

### 2.2. Global optimization

Several global optimization approaches have been effectively used in interpretation of different geophysical data (e.g., [35–37, 75, 77, 82–86]). In the present study, an alternative method

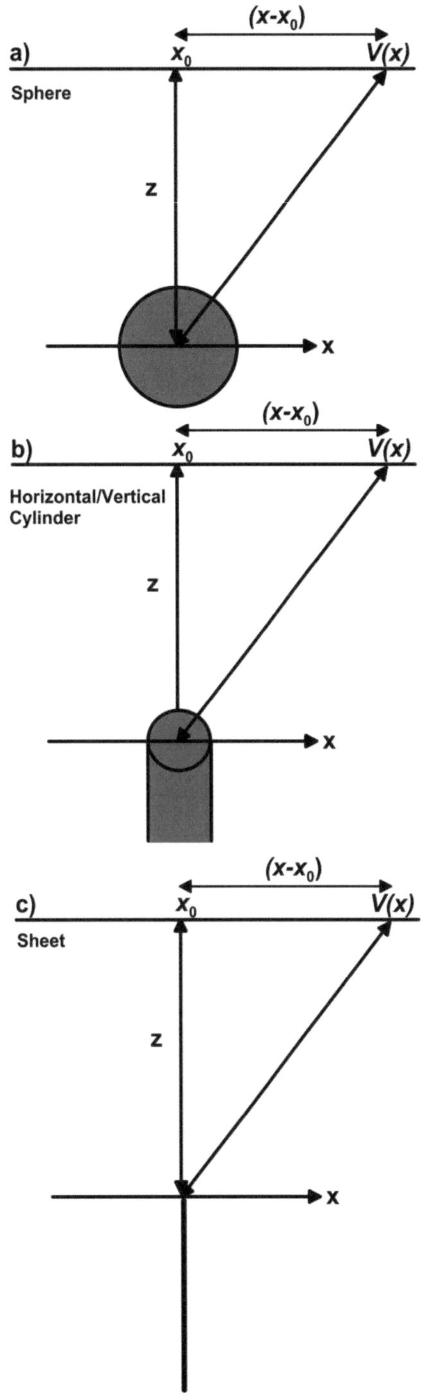


**Figure 1.** A diagram showing cross-sectional views, geometries, and parameters for (a) sphere, (b) cylinder, and (c) sheet-type structure.

of simulated annealing (SA) which is called as very fast simulated annealing (VFSA) was applied in the present study to elucidate the total gradient SP anomaly. VFSA is different from SA in terms of faster cooling schedule due to its finer Cauchy probability distribution for the arbitrary selection of every model parameter. VFSA takes any value in a model space, whereas SA does it in a predefined model space and hence the resolution increases for VFSA. Moreover, in VFSA optimization process, it does not recall all models and hence needs very small memory [35, 87, 88]. In every geophysical inversion, the main objective is to minimize the error function or the misfit. In the present work, the misfit (φ) between the observed and calculated/model response was used for SP data interpretation because of the fact that the objective function gets affected (increases) near zero crossing of SP anomaly (after [78]).

$$\phi = \frac{1}{N}\sum_{i=1}^{N}\left(\frac{v_i^0 - v_i^c}{\left|v_i^0\right| + \left(v_{\max}^0 - v_{\min}^0\right)/2}\right)^2 \tag{4}$$

where N is the number of data point, $V_i^0$ and $V_i^c$ are the $i$th observed and model responses and $V_{max}^0$ and $V_{min}^0$ are the maximum and minimum values of the observed response, respectively.

For the present inversion of SP data using VFSA optimization process, different parameters such as initial temperature, cooling schedule, number of iterations, and number of moves per temperature were taken as 1.0, 0.4, 2000, and 50. To find out the global/optimum solution, probability density function (PDF) was taken within 60.65% limit and ambiguity study has also been carried out based on the techniques developed by Mosegaard and Tarantola [89], Sen and Stoffa [90]. The details of the inversion process can be found in different literatures such as Sen and Stoffa [77], Sharma [86], Sharma and Biswas [78], Biswas [82], and Biswas [5]. Because in the earlier studies, the VFSA algorithm has established a very good performance as a powerful optimization method for estimating multidimensional and multimodal error functions. Hence, this optimization strategy is applied for SP data enhancement and regularization. The present VFSA algorithm for interpretation of ASA or TG of SP anomaly was carried out in Windows 8 environment using MS FORTRAN Developer studio on a simple desktop PC with Intel Core *i7* processor. For each step of optimization, an overall of $10^6$ forward computations (2000 iteration × 50 number of moves × 10 VFSA runs) were accomplished and accepted models were stored in memory. The total time required (not CPU time) to compute a sole inversion is 35 s.

## 3. Results and discussion

### 3.1. Parameter search range

An appropriate selection of initial guess values or the search range for every model parameters is a significant objective of any inversion approach. For the quantitative elucidation of SP anomaly from ASA, the horizontal position of the source is found from the highest anomaly, the depth from half width [71], and the size of the amplitude from highest amplitude and depth. Srivastava and Agarwal [70] stated that the most stable parameter for interpretation of

SP anomaly is the horizontal location, which changes very little. Moreover, it was discussed earlier that limiting the shape factor also gives the consistent results in terms of ambiguity and error [35]. Hence, in order to find the accurate or the true value of each model parameter, initially the search ranges for each model parameters were kept wide so as to find out the most probable solution. Next, the search spaces were reduced to find the more appropriate results. When the model parameter gives the utmost results and very least error, the location ($x_0$) and the shape factor ($q$) were fixed to their real/true values to further reduce the error in the amplitude and the depth so as to find the actual depth of the subsurface bodies and reduce ambiguity in the final interpretation.

### 3.2. Synthetic examples

The VFSA inversion technique was utilized considering synthetic noise-free and noise-corrupted data (10 and 20% random noise) for self-potential anomaly for various subsurface structures derived from the ASA anomaly. At first, every model parameter was inverted for every data. Synthetic hypothetical data were produced utilizing Eq. (2) for different idealized geobodies, and 10 and 20% random noise is corrupted to the synthetic data. VFSA inversion was applied utilizing noise-free and noisy data to retrieve the genuine model parameters and analyze the impact of noise on the deciphered model parameters. In general, a reasonable search range/space for every model parameter was chosen and one VFSA run was performed. Thereafter, the best possible convergence of every model parameter was studied ($k$, $x_0$, $z$, and $q$) and misfit.

Next, to acquire the mean model, 10 VFSA runs were executed. At that point, histograms were constructed from the retrieved models whose misfit error was lesser than $10^{-4}$. Then, a statistical mean model was calculated utilizing models that have misfit lesser than $10^{-4}$, which also exist inside one standard deviation. Besides, cross-plots were also developed and analyzed to check whether the model parameters were inside the high PDF region (60.65%). Further, noises were also added in the data and the process was repeated again where the misfit error was lower than $10^{-2}$. Finally, the comparison between the observed and model responses was shown for every model. This process was applied for each hypothetical synthetic noise-free, noisy, and field cases.

#### *3.2.1. Model 1 (sphere)*

Inversion of the SP data was performed as said above utilizing synthetic noise-free data for sphere-like structure. **Figure 2** demonstrates the convergence example for every model parameter. **Figure 3a** demonstrates the histogram for every model parameter ($k$, $x_0$, and $z$). The histogram uncovers that all the parameters of the body can be very much interpretable after inversion. Moreover, very less ambiguity in the interpretation of the main three model parameters was found. The cross-plots analysis (**Figure 4a**) demonstrated that the model parameters were near their true value (green). The final estimated model parameters were within the ambiguity limits and inside the high PDF region (red). The fittings between the observed and model response are shown in **Figure 6a**. The deduced model parameters and mean model are shown in **Table 1**. Next, another model was also selected to see the variation in the amplitude, location, and depth. Inversion was repeated the same way as discussed above. **Figure 3b**

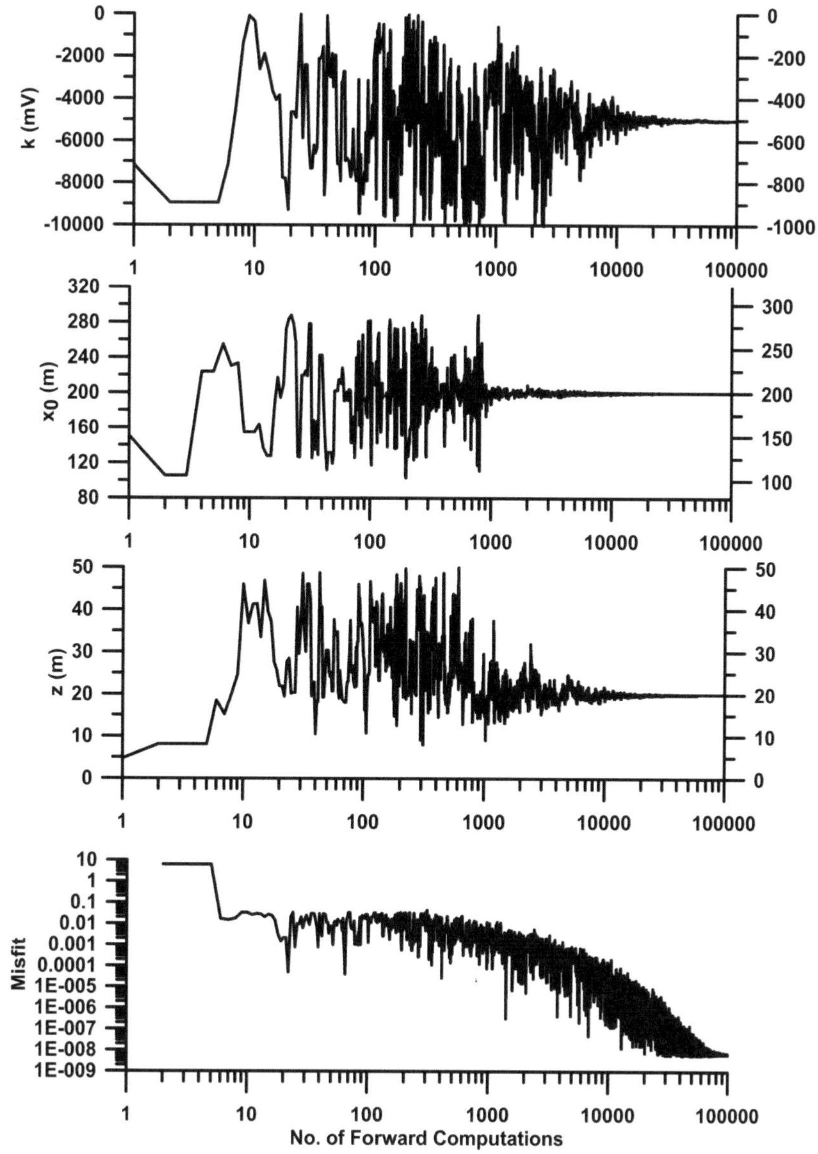


**Figure 2.** Convergence pattern for various model parameters and misfit.

shows the histogram for the three model parameters. **Figure 5a** demonstrates the cross-plots for this model as well. The fittings between the observed and model response are shown in **Figure 6b**, and the elucidated parameters and mean model are shown in **Table 2**.

### *3.2.2. Model 2 (horizontal and vertical cylinder)*

Another synthetic example for horizontal and vertical cylinder-like body was taken for inversion of the SP data. **Figure 7a** shows the histogram for every model parameter ($k$, $x_0$, and $z$).

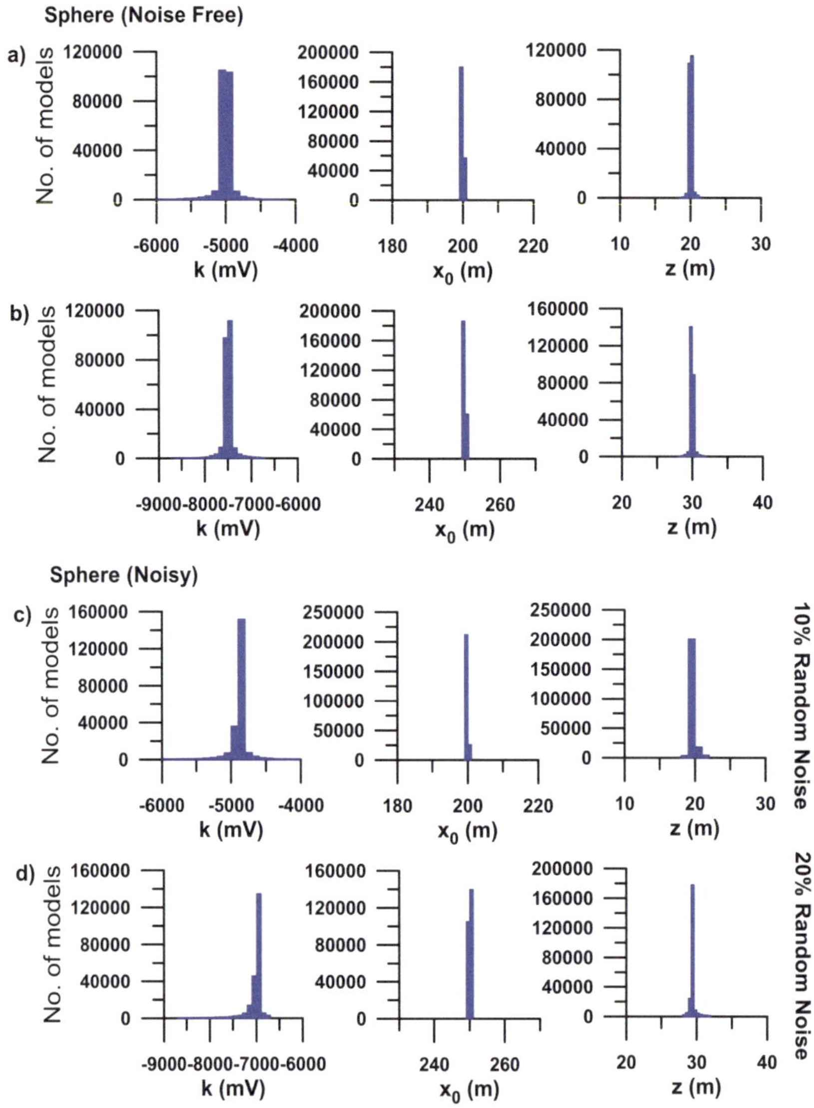


**Figure 3.** (a and b) Histograms of all accepted models having misfit <10–4 for noise-free synthetic data for sphere and (c and d) histograms of all accepted models having misfit <10–2 for noisy synthetic data (10 and 20% random noise) for sphere.

The histogram shows that all the parameters of the cylindrical structure can be very much elucidated. The cross-plots analysis (**Figure 8a**) demonstrates that the model parameters were near their true value (green) and the final expected model parameters were within the ambiguity limits and within high PDF region (red). Observed and model response fits are shown in **Figure 10a**. The elucidated parameters and mean model are shown in **Table 3**. Similarly, for the sphere-like structure, another model was selected to see the variation in the model parameters. **Figure 7b** demonstrates the histogram for the three model parameters. **Figure 9a**

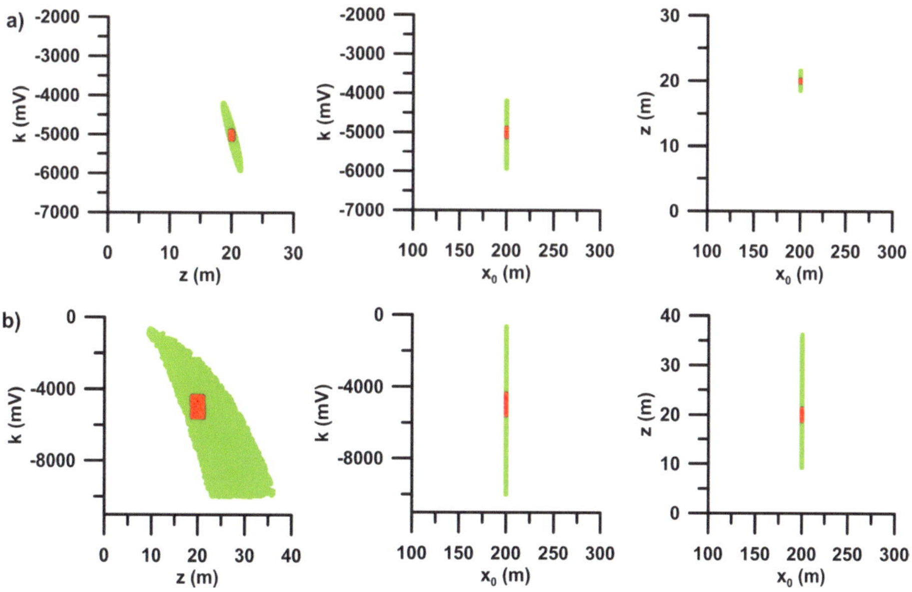


**Figure 4.** (a) Cross-plots between amplitude coefficient (*k*), depth (z), shape location ($x_0$) for all models having misfit < threshold ($10^{-4}$ for noise-free data) (green), and models with PDF > 60.65% (red) for noise-free data, (b) cross-plots between amplitude coefficient (*k*), depth (z), shape location ($x_0$) for all models having misfit < threshold ($10^{-2}$ for noisy data) (green), and models with PDF > 60.65% (red) for noisy (10% random) data for sphere.

| Model parameters | Actual value | Search range | Mean model (noise-free) | Mean model (noisy data) |
|---|---|---|---|---|
| *k* (mV) | −5000 | −10,000 to 0 | −5002.2 ± 27.4 | −4907.7 ± 152.2 |
| $x_0$ (m) | 200 | 0–300 | 200.0 ± 0.0 | 200.0 ± 0.0 |
| *z* (m) | 20 | 0–50 | 20.0 ± 0.0 | 19.9 ± 0.3 |
| *q* | 1.5 | 0–2 | 1.5 ± 0.0 | 1.5 ± 0.0 |
| Misfit | | | $6.6 \times 10^{-9}$ | $9.4 \times 10^{-5}$ |

**Table 1.** Actual model parameters, search range, and interpreted mean model for noise-free, 10% random noise with uncertainty for sphere.

demonstrates the cross-plots of this model. Observed and model responses are shown in **Figure 10b**, and the elucidated parameters and mean model are shown in **Table 4**.

### *3.2.3. Model 3 (inclined sheet)*

A synthetic example for 2-D inclined sheet-type body was also taken for inversion of the SP data. **Figure 11a** shows the histogram for every model parameter (*k*, $x_0$, and *z*). The cross-plots (**Figure 12a**) also demonstrate that the model parameters are near their actual value (green) and the final plausible model parameters lie inside the estimated ambiguity limits and inside high PDF region (red). Observed and model response fits are shown in **Figure 14a**. The elucidated

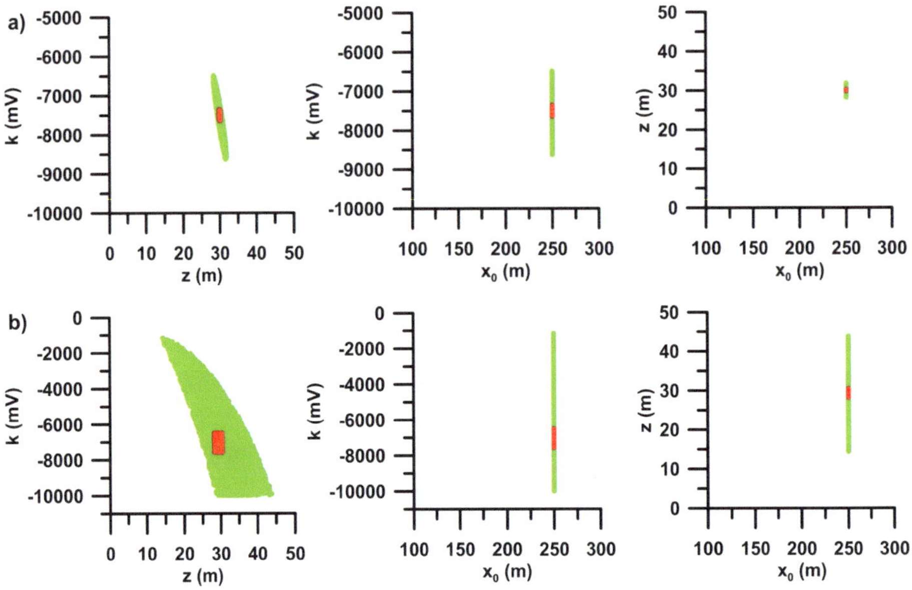


**Figure 5.** (a) Cross-plots between amplitude coefficient (*k*), depth (z), shape location ($x_0$) for all models having misfit < threshold ($10^{-4}$ for noise-free data) (green), and models with PDF > 60.65% (red) for noise-free data, (b) cross-plots between amplitude coefficient (*k*), depth (z), shape location ($x_0$) for all models having misfit < threshold ($10^{-2}$ for noisy data) (green), and models with PDF > 60.65% (red) for noisy (20% random) data for sphere.

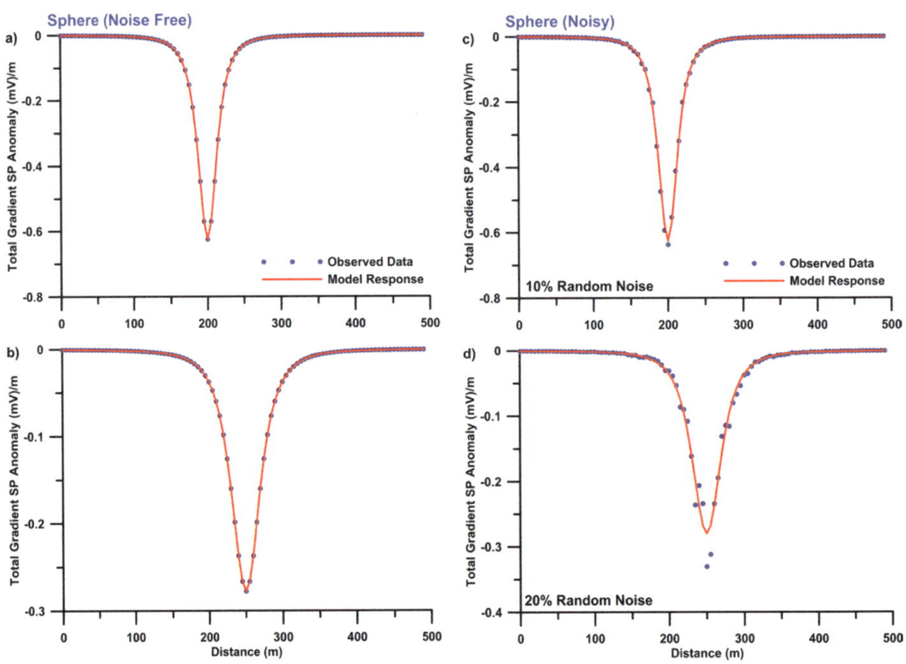


**Figure 6.** Fittings between the observed and model responses for sphere—(a) noise-free synthetic data; (b) noise-free synthetic data; (c) 10% random noisy synthetic data; and (d) 20% random noisy synthetic data.

| Model parameters | Actual value | Search range | Mean model (noise-free) | Mean model (noisy data) |
|---|---|---|---|---|
| $k$ (mV) | −7500 | −10,000 to 0 | −7497.4 ± 34.6 | −7037.4 ± 151.6 |
| $x_0$ (m) | 250 | 0–300 | 250.0 ± 0.0 | 250.0 ± 0.0 |
| $z$ (m) | 30 | 0–50 | 30.0 ± 0.0 | 29.9 ± 0.3 |
| $q$ | 1.5 | 0–2 | 1.5 ± 0.0 | 1.5 ± 0.0 |
| Misfit | | | $4.2 \times 10^{-8}$ | $7.1 \times 10^{-4}$ |

**Table 2.** Actual model parameters, search range, and interpreted mean model for noise-free, 20% random noise with uncertainty for sphere.

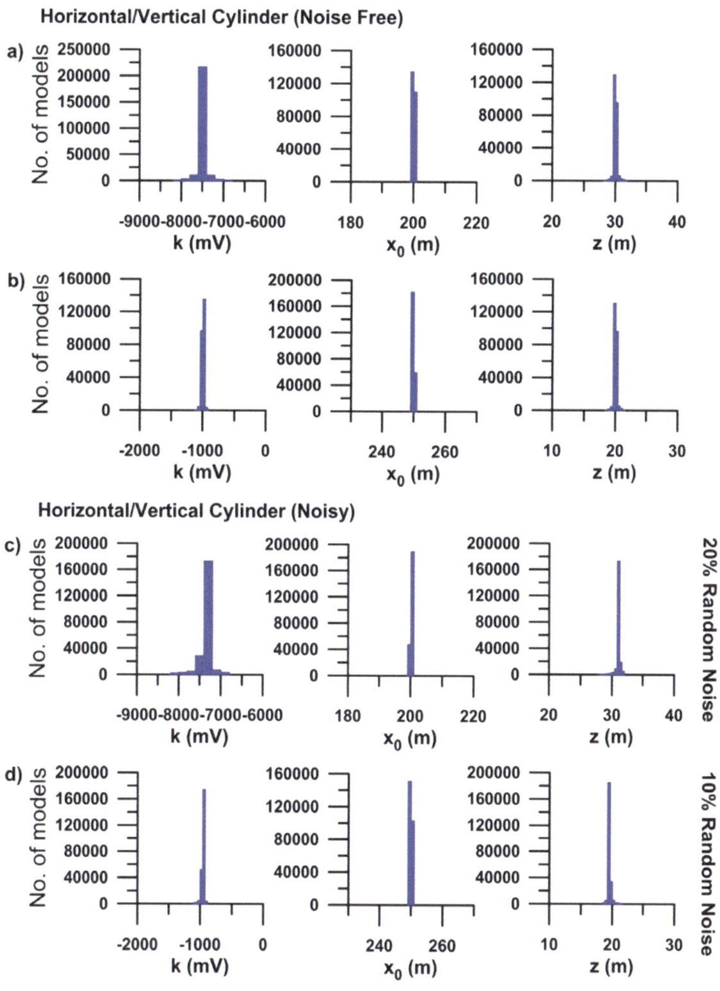


**Figure 7.** (a and b) Histograms of all accepted models having misfit < $10^{-4}$ for noise-free synthetic data for cylinder and (c and d) histograms of all accepted models having misfit < $10^{-2}$ for noisy synthetic data (20 and 10% random noise) for cylinder.

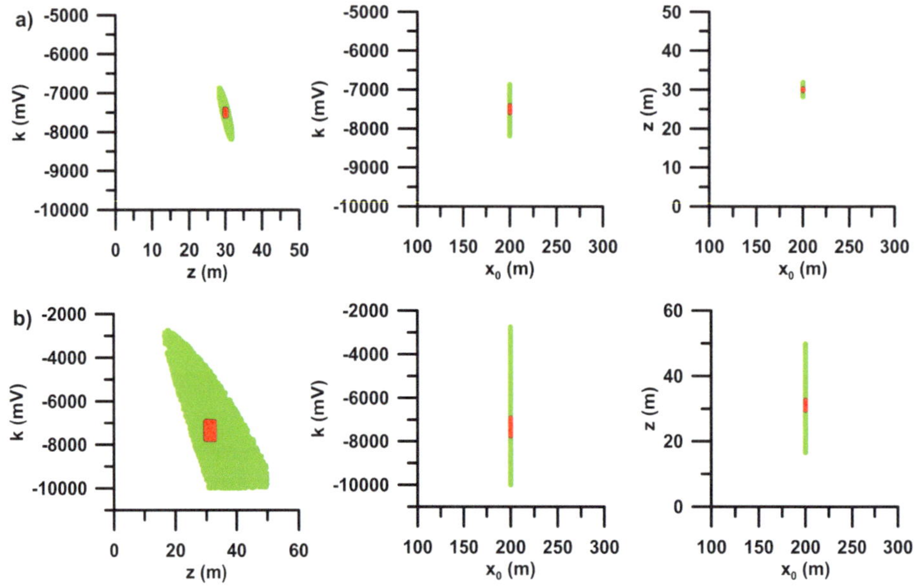


**Figure 8.** (a) Cross-plots between amplitude coefficient ($k$), depth (z), shape location ($x_0$) for all models having misfit < threshold ($10^{-4}$ for noise-free data) (green), and models with PDF > 60.65% (red) for noise-free data, (b) cross-plots between amplitude coefficient ($k$), depth (z), shape location ($x_0$) for all models having misfit < threshold ($10^{-2}$ for noisy data) (green), and models with PDF > 60.65% (red) for noisy (10% random) data for cylinder.

| Model parameters | Actual value | Search range | Mean model (noise-free) | Mean model (noisy data) |
|---|---|---|---|---|
| $k$ (mV) | −7500 | −10,000 to 0 | −7497.9 ± 23.7 | −7357.3 ± 122.8 |
| $x_0$ (m) | 200 | 0–300 | 200.0 ± 0.0 | 200.0 ± 0.0 |
| $z$ (m) | 30 | 0–50 | 30.0 ± 0.0 | 31.1 ± 0.4 |
| $q$ | 1.0 | 0–2 | 1.0 ± 0.0 | 1.0 ± 0.0 |
| Misfit | | | $5.2 \times 10^{-11}$ | $9.8 \times 10^{-4}$ |

**Table 3.** Actual model parameters, search range, and interpreted mean model for noise-free, 20% random noise with uncertainty for horizontal/vertical cylinder.

parameters and mean model are given in **Table 5**. Another model was selected to see the variation in the model parameters. **Figure 11b** displays the histogram for the three model parameters. **Figure 13a** shows the cross-plots of this model. Observed and model response fits are shown in **Figure 14b**, and the elucidated parameters and mean model are shown in **Table 6**.

### 3.3. Noise analysis

To see the effectiveness of the present inversion results, the synthetic noise-free data must be corrupted with different degrees of noises and reinterpreted using the inversion method to check its performance and robustness of the method. Hence, 10 and 20% random noise was added to the data for sphere, cylinder, and sheet-type structure, and the procedure was repeated again to examine the effect of noise. **Figure 3c** and **d** displays the histogram for noise-corrupted

http://dx.doi.org/10.5772/intechopen.79111

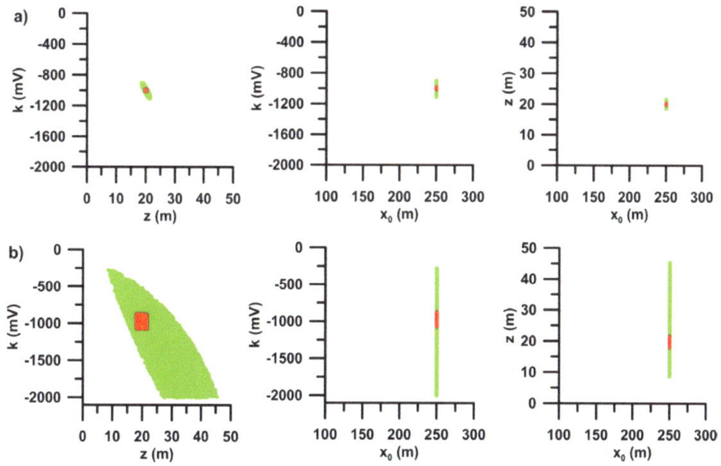


**Figure 9.** (a) Cross-plots between amplitude coefficient (*k*), depth (z), shape location ($x_0$) for all models having misfit < threshold ($10^{-4}$ for noise-free data) (green), and models with PDF > 60.65% (red) for noise-free data, (b) cross-plots between amplitude coefficient (*k*), depth (z), shape location ($x_0$) for all models having misfit < threshold ($10^{-2}$ for noisy data) (green), and models with PDF > 60.65% (red) for noisy (20% random) data for cylinder.

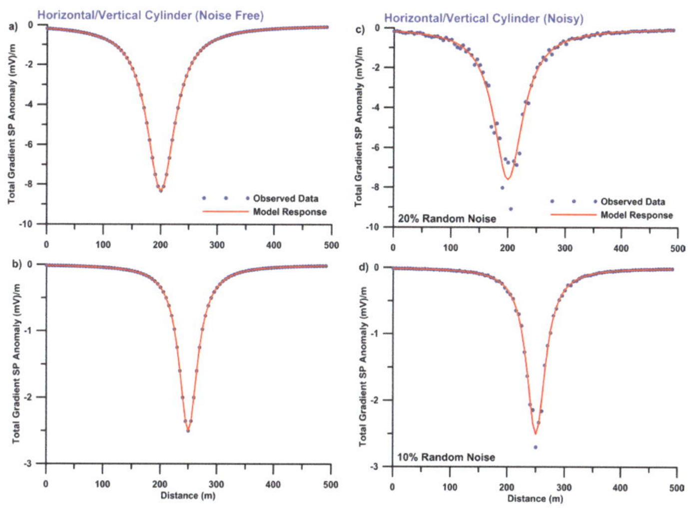


**Figure 10.** Fittings between the observed and model responses for cylinder—(a) noise-free synthetic data; (b) noise-free synthetic data; (c) 10% random noisy synthetic data; and (d) 20% random noisy synthetic data.

sphere-like structure. **Figures 4b** and **5b** show the cross-plots for noisy data. **Tables 1** and **2** show the interpreted mean model for noisy data and the fits between the observed and model response for noisy model are shown in **Figure 6c** and **d**. For horizontal and vertical cylinder-like structure, again noise was corrupted with the two models mentioned above. **Figure 7c** and **d** shows the histogram for noise-corrupted horizontal/vertical-like structure. **Figures 8b** and **9b** show the cross-plots for noisy data. **Tables 3** and **4** show the interpreted mean model for noisy data and the fits between the observed and model response for noisy model are shown in **Figure 10c** and **d**. Also, for 2-D inclined sheet-type structure, the same amount of noise was

| Model parameters | Actual value | Search range | Mean model (noise-free) | Mean model (noisy data) |
|---|---|---|---|---|
| $k$ (mV) | −1000 | −2000 to 0 | −999.6 ± 3.3 | −963.1 ± 22.6 |
| $x_0$ (m) | 250 | 0–300 | 250.0 ± 0.0 | 250.0 ± 0.0 |
| $z$ (m) | 20 | 0–50 | 20.0 ± 0.0 | 19.6 ± 0.4 |
| $q$ | 1.0 | 0–2 | 1.0 ± 0.0 | 1.0 ± 0.0 |
| Misfit | | | $5.9 \times 10^{-9}$ | $1.6 \times 10^{-4}$ |

**Table 4.** Actual model parameters, search range, and interpreted mean model for noise-free, 10% random noise with uncertainty for horizontal/vertical cylinder.

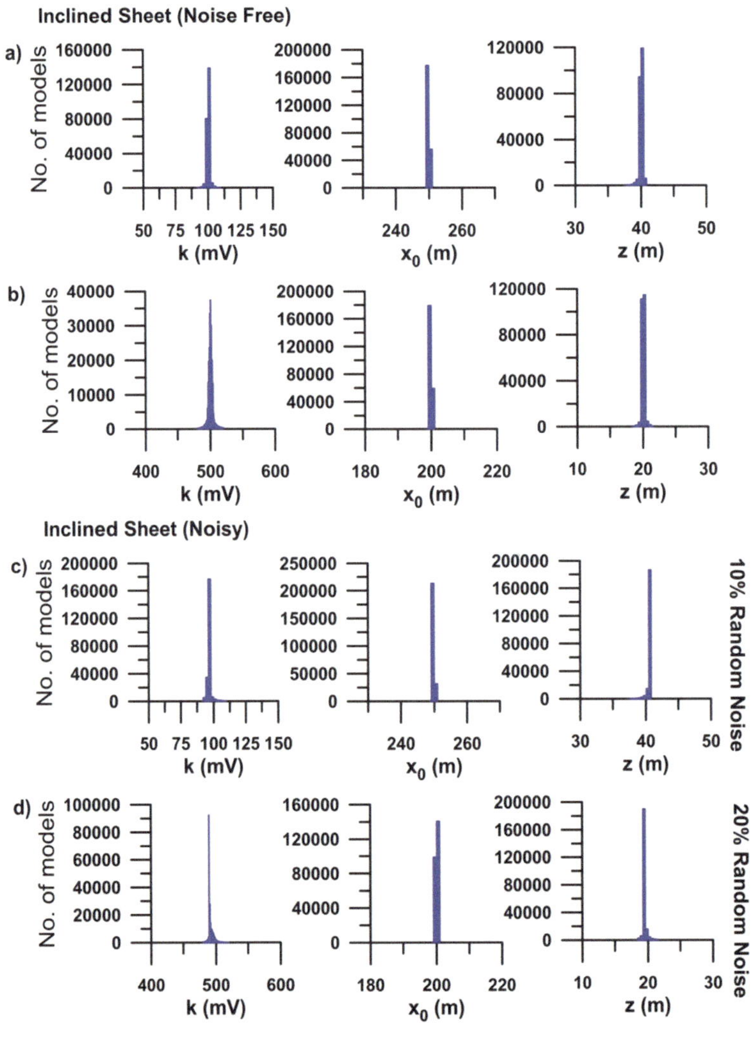


**Figure 11.** (a and b) Histograms of all accepted models having misfit < $10^{-4}$ for noise-free synthetic data for cylinder and (c, d) histograms of all accepted models having misfit < $10^{-2}$ for noisy synthetic data (10 and 20% random noise) for sheet.

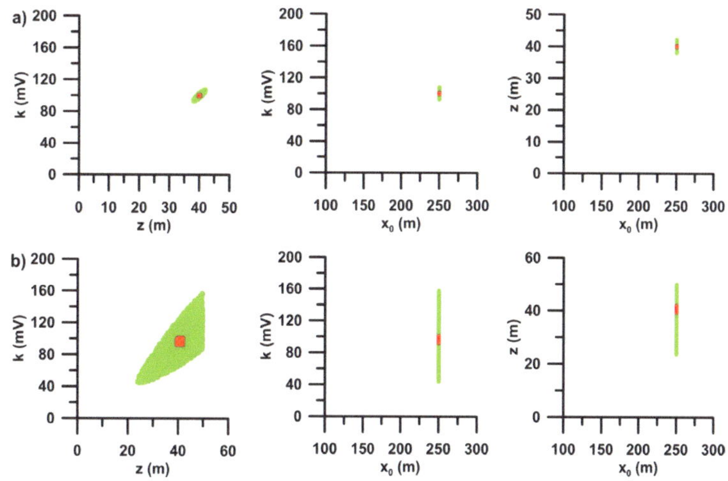


**Figure 12.** (a) Cross-plots between amplitude coefficient (*k*), depth (z), shape location ($x_0$) for all models having misfit < threshold ($10^{-4}$ for noise-free data) (green), and models with PDF > 60.65% (red) for noise-free data, (b) cross-plots between amplitude coefficient (*k*), depth (z), shape location ($x_0$) for all models having misfit < threshold ($10^{-2}$ for noisy data) (green), and models with PDF > 60.65% (red) for noisy (10% random) data for sheet.

| Model parameters | Actual value | Search range | Mean model (noise-free) | Mean model (noisy data) |
|---|---|---|---|---|
| *k* (mV) | 100 | 0–200 | 100.1 ± 0.4 | 96.3 ± 1.2 |
| $x_0$ (m) | 250 | 0–300 | 250.0 ± 0.0 | 250.0 ± 0.0 |
| *z* (m) | 40 | 0–50 | 40.0 ± 0.0 | 40.6 ± 0.4 |
| *q* | 1.0 | 0–2 | 1.0 ± 0.0 | 1.0 ± 0.0 |
| Misfit | | | $4.7 \times 10^{-7}$ | $8.1 \times 10^{-5}$ |

**Table 5.** Actual model parameters, search range, and interpreted mean model for noise-free, 10% random noise with uncertainty for sheet.

corrupted in the models mentioned above. **Figure 11c** and **d** shows the histogram for noise-corrupted inclined sheet-type structure. **Figures 12b** and **13b** illustrate the cross-plots for noisy data. **Tables 5** and **6** show the interpreted mean model for noisy data and the fits between the observed and model response for noisy model are shown in **Figure 14c** and **d**. It can be seen from the study of histogram and cross-plots that the model parameters are interpreted very precisely, and it also advocates that the appraised model parameters for all structures are inside the ambiguity limits and within high PDF region. This also suggests that the inversion methodology developed for the elucidation of SP anomalies can precisely decide every model parameter and even if the data are highly corrupted with different degrees of noises.

### 3.4. Effect of complicated structure

It is significant to mention that that, in nature (field examples), it is very difficult to get an idealized geobody or structure. Moreover, the structures are mostly corrupted with different

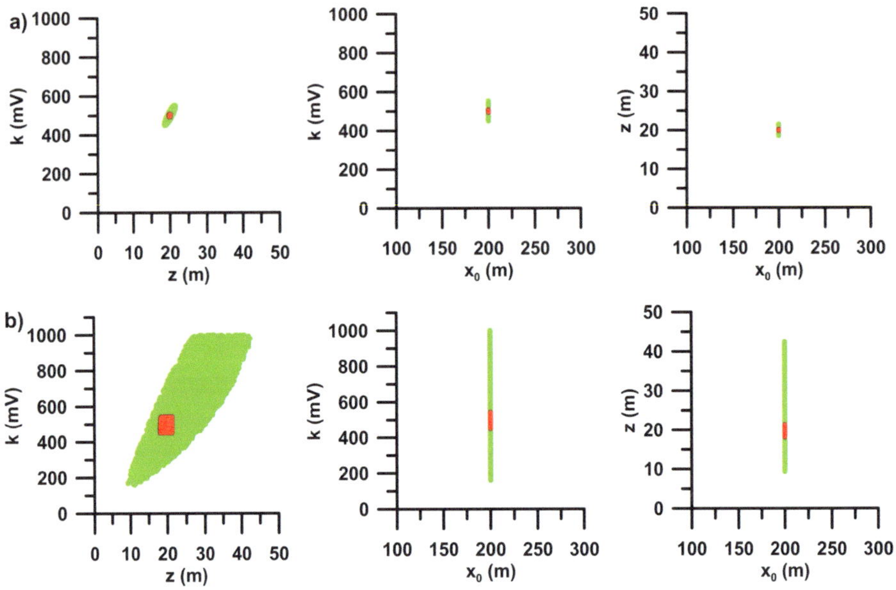


**Figure 13.** (a) Cross-plots between amplitude coefficient ($k$), depth (z), shape location ($x_0$) for all models having misfit < threshold ($10^{-4}$ for noise-free data) (green), and models with PDF > 60.65% (red) for noise-free data, (b) cross-plots between amplitude coefficient ($k$), depth (z), shape location ($x_0$) for all models having misfit < threshold ($10^{-2}$ for noisy data) (green), and models with PDF > 60.65% (red) for noisy (20% random) data for sheet.

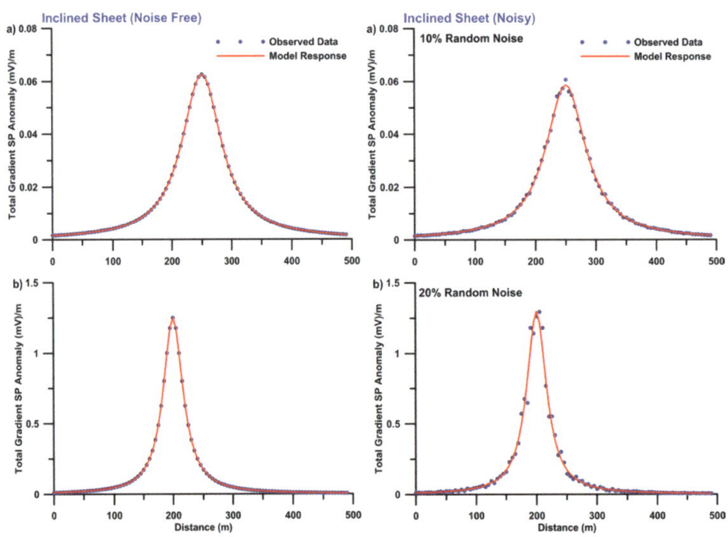


**Figure 14.** Fittings between the observed and model responses for sheet—(a) noise-free synthetic data; (b) noise-free synthetic data; (c) 10% random noisy synthetic data; and (d) 20% random noisy synthetic data.

degree of noises. Also, there might be multiple structures within the subsurface of different types. In those cases, the anomaly from multiple structures will be very difficult to interpret. To test whether the inversion method can accurately identify the multiple structures, three different structures were taken and forward responses have been computed using Eqs. (2)

| Model parameters | Actual value | Search range | Mean model (noise-free) | Mean model (noisy data) |
|---|---|---|---|---|
| $k$ (mV) | 500 | 0–1000 | 500.0 ± 1.6 | 493.5 ± 10.5 |
| $x_0$ (m) | 200 | 0–300 | 200.0 ± 0.0 | 200.0 ± 0.0 |
| $z$ (m) | 20 | 0–50 | 20.0 ± 0.0 | 19.5 ± 0.3 |
| $q$ | 1.0 | 0–2 | 1.0 ± 0.0 | 1.0 ± 0.0 |
| Misfit | | | $2.0 \times 10^{-9}$ | $6.2 \times 10^{-4}$ |

**Table 6.** Actual model parameters, search range, and interpreted mean model for noise-free, 20% random noise with uncertainty for sheet.

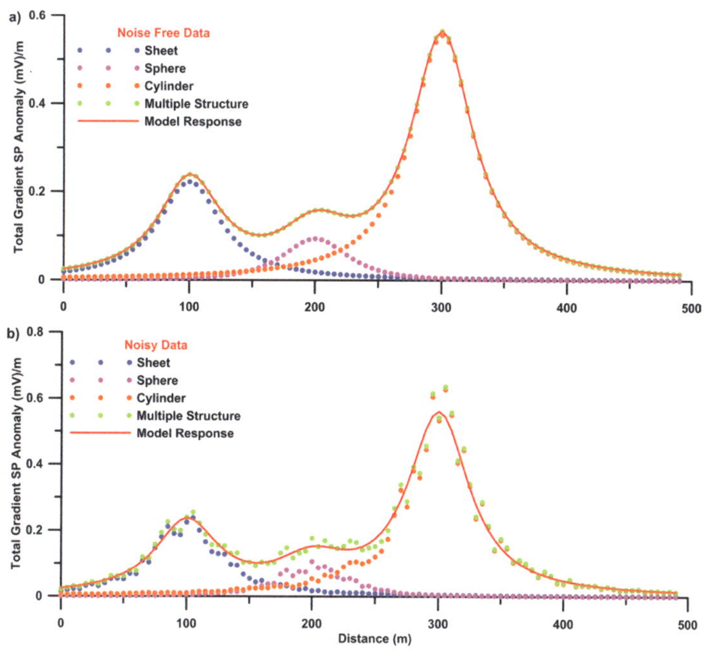


**Figure 15.** Fittings between the observed and model responses for multiple structure—(a) noise-free synthetic data; (b) 20% random noisy synthetic data.

and (3). Single structure and multiple structures from the forward modeling are shown in **Figure 15a**. Inversion of the synthetic data was carried out the same way as for the single model examples. Fits between the observed and model response are also shown in **Figure 15a**. Moreover, to investigate the effect of noisy data, only 20% random noise was corrupted in the data and inversion was repeated. **Figure 15b** displays the fits between the observed and model response for noisy data. **Table 7** shows the interpreted model parameters for each structure derived from the multiple anomalies. Both noise-free and noisy data are shown in **Table 7**. It can be seen from **Table 7** that the misfit is quite less for noise-free and noise-corrupted data. Histogram and cross-plots show alike as revealed in other examples for single structures. However, it is not presented here for brevity.

| Model parameters | Actual value | Search range | Sheet | Actual value | Search range | Sphere | Actual value | Search range | H/V cylinder |
|---|---|---|---|---|---|---|---|---|---|
| *Multiple bodies—noise-free* | | | | | | | | | |
| $k$ (mV) | 200 | 0–500 | 200.2 ± 0.6 | 6000 | 1000–10,000 | 6003.5 ± 26.2 | 500 | 100–1000 | 500.3 ± 1.6 |
| $x_0$ (m) | 100 | 0–200 | 100.0 ± 0.0 | 200 | 0–300 | 200.0 ± 0.0 | 300 | 50–200 | 300.0 ± 0.0 |
| $z$ (m) | 30 | 0–50 | 30.0 ± 0.1 | 40 | 0–50 | 40.0 ± 0.0 | 30 | 0–100 | 30.0 ± 0.1 |
| $q$ | 1.0 | 0–2 | 1.0 ± 0.0 | 1.5 | 0–2 | 1.5 ± 0.0 | 1.0 | 0–2 | 1.0 ± 0.0 |
| Misfit | | | $7.4 \times 10^{-8}$ | | | $3.3 \times 10^{-7}$ | | | $1.6 \times 10^{-8}$ |
| *Multiple bodies—20% random noise* | | | | | | | | | |
| $k$ (mV) | 200 | 0–500 | 205.4 ± 4.3 | 6000 | 1000–10,000 | 6316.9 ± 133.6 | 500 | 100–1000 | 485.4 ± 11.0 |
| $x_0$ (m) | 100 | 0–200 | 100.0 ± 0.0 | 200 | 0–300 | 200.0 ± 0.0 | 300 | 50–200 | 300.0 ± 0.0 |
| $z$ (m) | 30 | 0–50 | 30.5 ± 0.5 | 40 | 0–50 | 41.4 ± 0.4 | 30 | 0–100 | 29.7 ± 0.5 |
| $q$ | 1.0 | 0–2 | 1.0 ± 0.0 | 1.5 | 0–2 | 1.5 ± 0.0 | 1.0 | 0–2 | 1.0 ± 0.0 |
| Misfit | | | $7.8 \times 10^{-4}$ | | | $9.9 \times 10^{-4}$ | | | $1.2 \times 10^{-3}$ |

**Table 7.** Actual model parameters, search range, and interpreted mean model for noise-free and 20% random noise with uncertainty for multiple structure.

## 3.5. Field example

To demonstrate the efficiency of the present method, two field examples from Turkey and Canada were taken for interpretation. The field anomaly of ASA/TG from SP anomaly was taken from the earlier published literatures as described below:

### *3.5.1. Sulleymonkoy anomaly, Ergani, Turkey*

This field example was taken from the Sulleymonkoy SP anomaly, Ergani, Turkey [91]. The ASA or TG anomaly was derived following Srivastava and Agarwal [70] (**Figure 18**). The anomaly was interpreted by many workers using different interpretation methods [38, 92–96]. The SP anomaly data were interpreted using the VFSA global optimization technique in this study. The TG anomaly shows a large magnitude peak anomaly with two small anomalies. In the present case, initially the main peak anomaly was interpreted considering a single structure. The depth obtained by the present study for single body was found to be 34.5 m on the horizontal location of 66.3 m, and the structure was interpreted as a cylindrical body. Histogram plot for single body (**Figure 16a**) also shows that the model parameters were precisely determined. Analysis of cross-plots (**Figure 17a**) also shows that the assessed model parameters were within the ambiguity limits. Next, considering multiple structures, the inversion process was repeated again for three different structures considering three peak values. Analysis of histogram plot for multiple bodies (**Figure 16b–d**) also shows that the model parameters were precisely determined. Investigation of cross-plots (**Figure 17b–d**) also shows that the estimated model parameters were within the uncertainty limits and

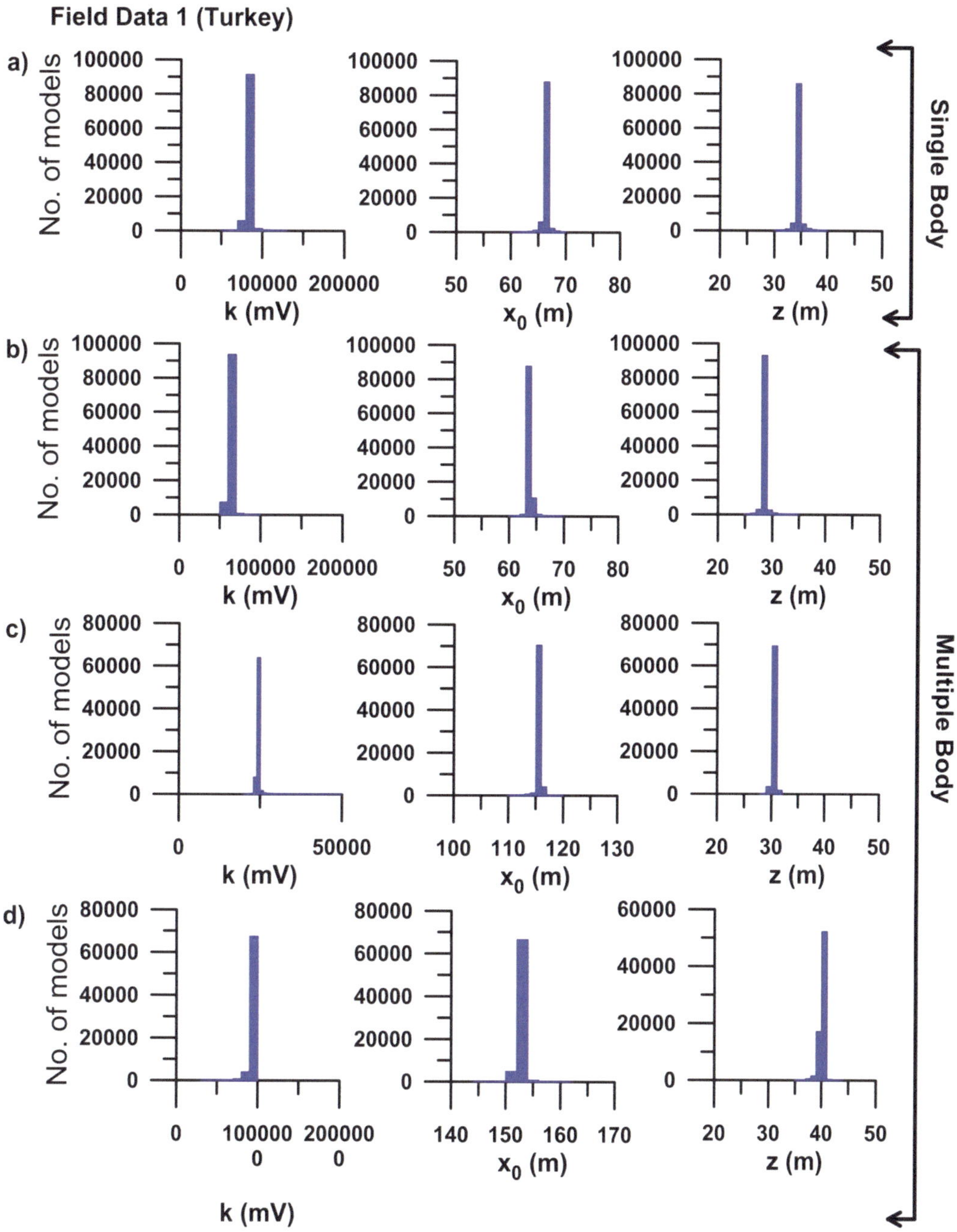


**Figure 16.** (a) Histograms of all accepted models having misfit < $10^{-2}$ for field data—single body and (b–d) histograms of all accepted models having misfit < $10^{-2}$ for field data—multiple bodies—Sulleymonkoy anomaly, Ergani, Turkey.

more precise than considering single body using the whole TG anomaly. However, the depth obtained for the peak anomaly was found to be 28.5 m on the location of 63.9 m considering multiple structures. The depth obtained by other workers such as Yungul [91] as 38.8 m, Bhattacharya and Roy [94] as 40.0 m, Agarwal [92] as 30.1 m, Sundararajan and Srinivas [96]

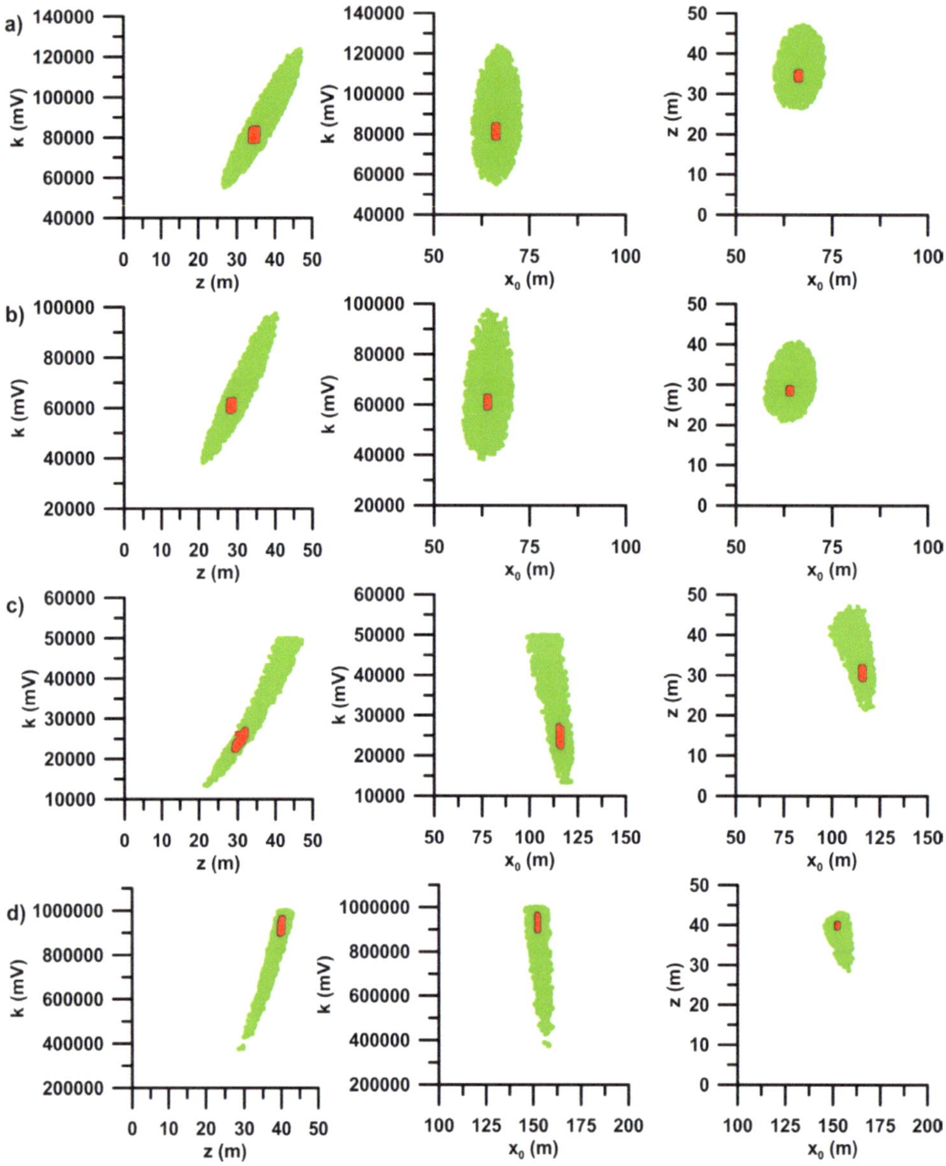


**Figure 17.** (a) Cross-plots between amplitude coefficient ($k$), depth (z), shape location ($x_0$) for all models having misfit < threshold ($10^{-2}$ for noise-free data) (green), and models with PDF > 60.65% (red) for field data—single body, (b–d) cross-plots between amplitude coefficient ($k$), depth (z), shape location ($x_0$) for all models having misfit < threshold ($10^{-2}$ for noise-free data) (green), and models with PDF > 60.65% (red) for field data—multiple bodies—Sulleymonkoy anomaly, Ergani, Turkey.

as 36 m, and Srivastava and Agarwal [95] as 28.9 m. Moreover, the estimated uncertainty was less considering multiple structures rather than a single body. **Table 8** shows a comparison between the different model parameters and misfit. **Figure 18** shows the comparison between the observed TG anomaly and the model response.

| Model parameters | Single body | | Multiple bodies | | | | | |
|---|---|---|---|---|---|---|---|---|
| | Search range | Present method B1 | Search range | Present method B1 | Search range | Present method B2 | Search range | Present method B3 |
| $k$ (mV) | 1000–200,000 | 81281.0 ± 957.5 | 1000–100,000 | 60888.2 ± 574.0 | 100–50,000 | 24244.9 ± 442.6 | 1000–1,000,000 | 935524.0 ± 12801.3 |
| $x_0$ (m) | 50–100 | 66.3 ± 0.2 | 50–100 | 63.9 ± 0.2 | 50–200 | 115.8 ± 0.2 | 50–200 | 152.3 ± 0.2 |
| $z$ (m) | 0–50 | 34.5 ± 0.3 | 0–50 | 28.5 ± 0.2 | 0–50 | 30.3 ± 0.3 | 0–100 | 40.0 ± 0.2 |
| $q$ | 0–2 | 1.0 | 0–2 | 1.0 | 0–2 | 1.0 | 0–2 | 1.0 |
| Misfit | | $5.6 \times 10^{-3}$ | | $9.8 \times 10^{-4}$ | | $8.5 \times 10^{-5}$ | | $1.4 \times 10^{-3}$ |

**Table 8.** Search range and interpreted mean model for Sulleymonkoy anomaly, Ergani, Turkey.

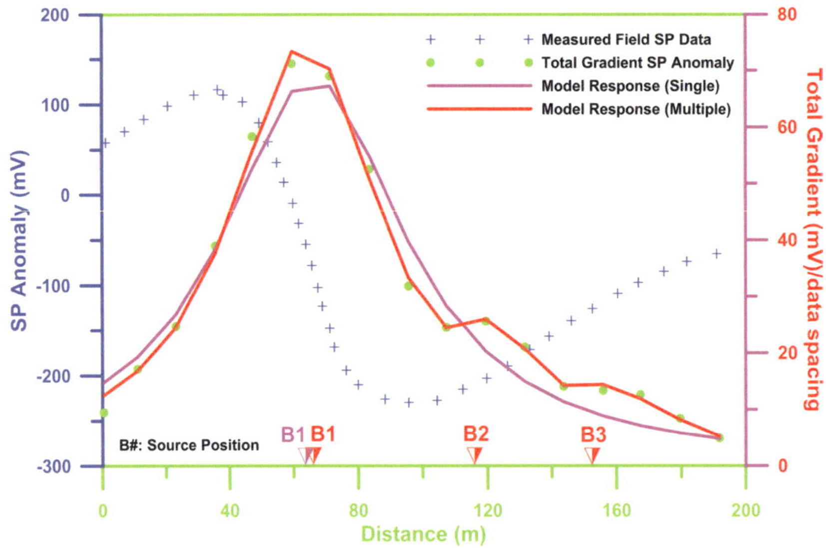


**Figure 18.** Fittings between the observed and model responses for Sulleymonkoy anomaly, Ergani, Turkey.

### 3.5.2. *Senneterre area of Quebec, Canada*

A total gradient of SP anomaly (**Figure 6a**) over a massive sulfide ore deposits in the Senneterre area of Quebec, Canada [97], was considered for this study. The anomaly was interpreted using enhanced local wave number technique by Srivastava and Agarwal [95] and regularized inversion by Mehanee [7] without considering TG. However, the anomaly was reinterpreted by Srivastava et al. [75] using ACO technique. Srivastava et al. [75] considered six anomalies from the TG. This anomaly was also elucidated using the present inversion method to retrieve the model parameters considering four peak anomalies, which were quite distinct. Investigation of histogram plot for multiple bodies (**Figure 19a–d**) also shows that the model parameters were precisely determined. Examination of cross-plots (**Figure 20a–d**) also shows that the appraised mean model parameters were within the ambiguity limits. The depth obtained by using the VFSA method was found to be 10.5, 4.6, 3.7, and 4.3 m, respectively. The depths obtained by Srivastava et al. [75] for the four bodies are 5, 4.2, 3.8, and 4.3 m, respectively. The assessed parameters in this study are in respectable agreement with the other work. Moreover, in the present work, all four anomalies were interpreted as multiple structures and not independently. The estimated model parameters are shown in **Table 9** along with misfit. A comparison between the observed anomaly and model responses are shown in **Figure 21**.

## 3.6. Conclusions

It is well known that the local search inversion or the optimization has faster convergence rate. However, while trying to search the global optima, it can be trapped in the local minima. Hence, selection of initial guess is very important for global optimization studies. In case of complicated structure, absence of a priori information also hampers the final solution. However, global optimization methods search the best possible solution and try to find out the exact solution

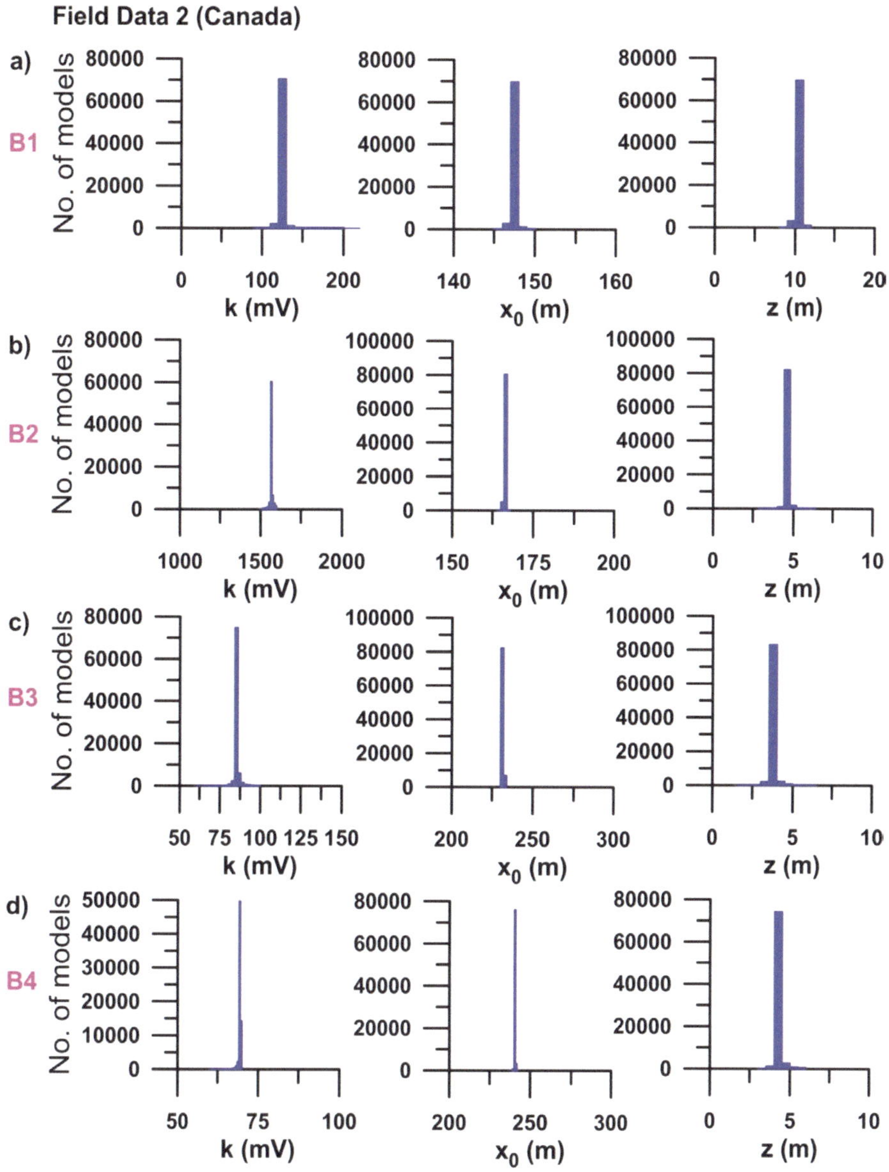


**Figure 19.** (a) Histograms of all accepted models having misfit < $10^{-2}$ for field data—single body and (b–d) histograms of all accepted models having misfit < $10^{-2}$ for field data—multiple bodies—Senneterre area of Quebec, Canada.

within a multidimensional model space. Moreover, apart from global optimization method, a statistical method can also be applied to find out the ambiguity associated with the final results.

In this study, an effort was made to examine the relevance and adequacy of VFSA on the parameter appraisals from ASA or TG of SP anomalies. In this method, the experimental studies were executed utilizing hypothetically derived data and field data. The elucidation of the amplitude coefficient, location, depth, and shape of a subsurface structure from TG

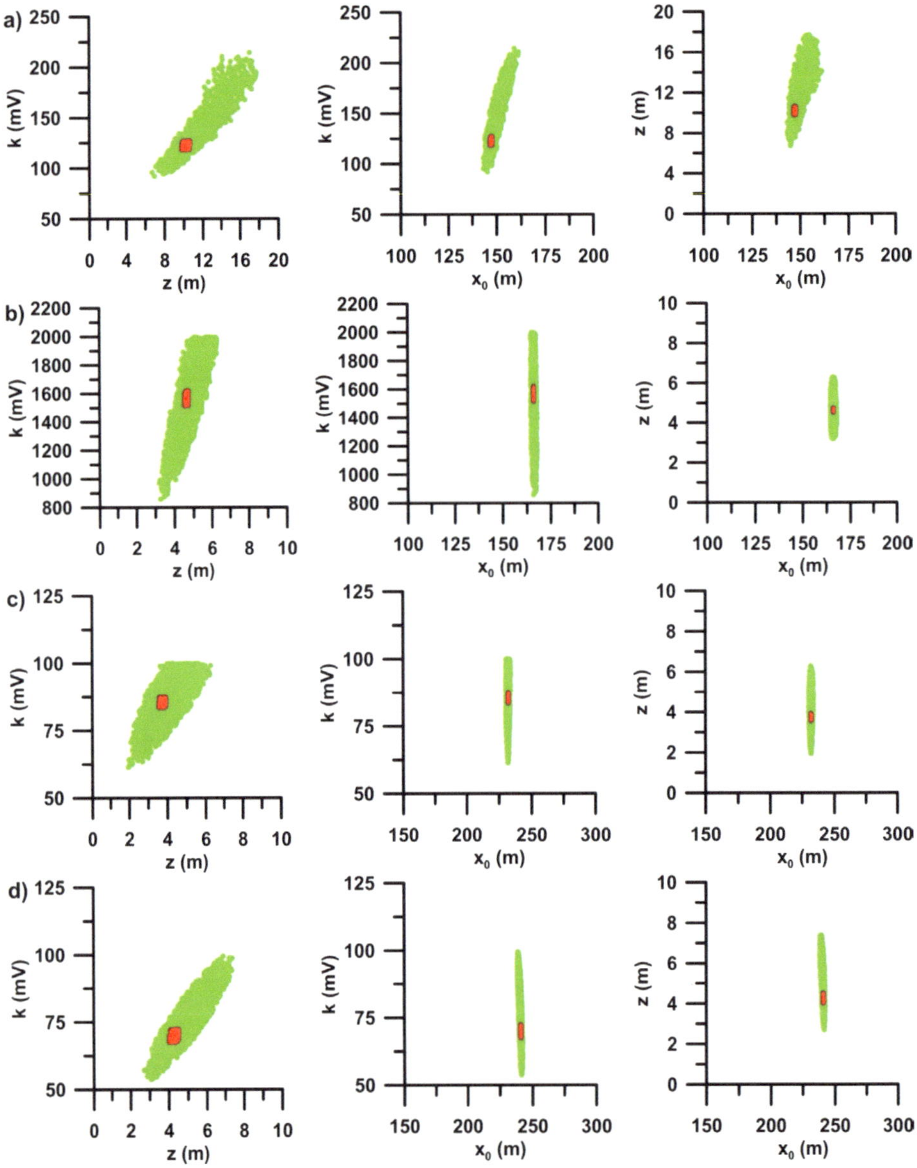


**Figure 20.** (a) Cross-plots between amplitude coefficient (*k*), depth (z), shape location ($x_0$) for all models having misfit < threshold ($10^{-2}$ for noise-free data) (green), and models with PDF > 60.65% (red) for field data—single body, (b–d) cross-plots between amplitude coefficient (*k*), depth (z), shape location ($x_0$) for all models having misfit < threshold ($10^{-2}$ for noise-free data) (green), and models with PDF > 60.65% (red) for field data—multiple bodies—Senneterre area of Quebec, Canada.

anomaly can be very much established utilizing the present technique. Synthetic data tests were performed utilizing both noise-free and noisy information sets derived from idealized geobodies. The present work reveals, while interpreting every single model parameter (amplitude coefficient, location, depth, shape) together, the VFSA method produces excellent results. Moreover, multiple model bodies were interpreted very efficiently.

http://dx.doi.org/10.5772/intechopen.79111

| Model parameters | Search range | Present method B1 | Search range | Present method B2 | Search range | Present method B3 | Search range | Present method B4 |
|---|---|---|---|---|---|---|---|---|
| $k$ (mV) | 10–500 | 122.1 ± 0.7 | 100–2000 | 1568.4 ± 13.9 | 0–100 | 85.4 ± 0.6 | 0–100 | 69.5 ± 0.3 |
| $x_0$ (m) | 100–200 | 147.2 ± 0.1 | 100–200 | 166.1 ± 0.1 | 200–250 | 232.0 ± 0.1 | 200–250 | 240.9 ± 0.1 |
| $z$ (m) | 0–20 | 10.5 ± 0.1 | 0–10 | 4.6 ± 0.1 | 0–10 | 3.7 ± 0.1 | 0–10 | 4.3 ± 0.1 |
| $q$ | 0–2 | 0.5 | 0–2 | 1.0 | 0–2 | 0.5 | 0–2 | 0.5 |
| Misfit | | $3.6 \times 10^{-3}$ | | $2.2 \times 10^{-3}$ | | $3.4 \times 10^{-3}$ | | $5.5 \times 10^{-3}$ |
| Srivastava et al. [75] (ACO) $x_0$ and $z_0$ are the coordinate of the source location in unit of data spacing (0.61 m). | | | | | | | | |
| | **Search space** | **TGM peak 1** | **Search space** | **TGM peak 3** | **Search space** | **TGM peak 5** | **Search space** | **TGM peak 6** |
| $k$ (mV) | 1–6250 | 157 ± 17 | 1–42,000 | 6998 ± 72 | 1–15,000 | 2201 ± 36 | 1–13,000 | 944 ± 11 |
| $x_0$ (m) | 220–240 | 143.2 ± 2.0 | 265–285 | 167.4 ± 2.1 | 370–390 | 231.2 ± 1.2 | 380–410 | 242.5 ± 3.1 |
| $z$ (m) | 1–50 | 5.0 ± 0.2 | 1–50 | 4.2 ± 0.2 | 1–50 | 3.8 ± 0.3 | 1–50 | 4.3 ± 0.1 |
| $\beta$ (=2$q$) | 0.2–4 | 1.3 ± 0.1 | 0.2–4 | 2.2 ± 0.3 | 0.2–4 | 2.2 ± 0.3 | 0.2–4 | 2.3 ± 0.3 |

**Table 9.** Search range and interpreted mean model for Senneterre area of Quebec, Canada.

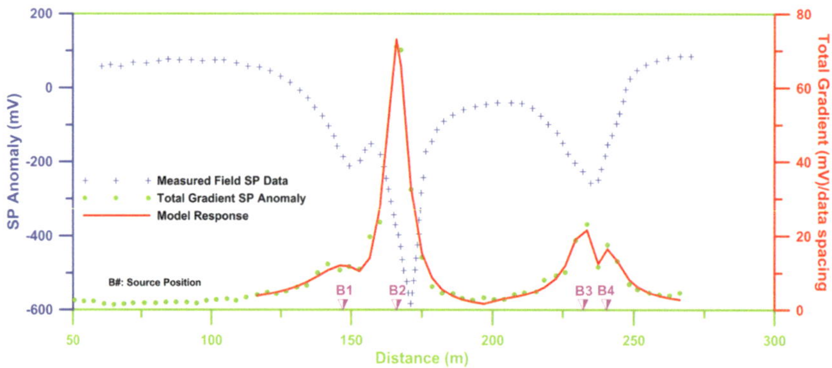


**Figure 21.** Fittings between the observed and model responses for Senneterre area of Quebec, Canada.

The subsequent histogram and cross-plots investigation proposes that the acquired parameters were within the high probability and less uncertainty. This is also supported by the rapid and stable convergence rate of the present inversion method. The viability of this approach has been effectively demonstrated employing noise-free and noisy data. The suitability of this technique for useful application in mineral investigation has been additionally shown on two field cases (Sulleymonkoy anomaly, Ergani, Turkey, Senneterre area of Quebec, Canada). The field data were presumed to be formed due to different idealized geological bodies as mentioned in the published literature. The technique can be utilized to understand numerous structures from the anomaly. The assessed inverse parameters for the field examples were observed to be in fair agreement with other alternate techniques such as PSO.

It is noteworthy to mention that the present work does not get affected by wide search range. Other studies suggested that the search range must be carried out by the workers to understand the effect of search range and the solution.

## Acknowledgements

I thank Prof. K.S. Essa for inviting me to submit a chapter for the edited book. Anonymous reviewer and the editor are thanked for their useful suggestions that have improved the quality of the present work.

## Author details

Arkoprovo Biswas

Address all correspondence to: arkoprovo@gmail.com

Department of Geology, Centre of Advanced Study, Institute of Science, Banaras Hindu University, Varanasi, UP, India

http://dx.doi.org/10.5772/intechopen.79111

## References

[1] Biswas A. A review on modeling, inversion and interpretation of self-potential in mineral exploration and tracing Paleo-shear zones. Ore Geology Reviews. 2017;**91**:21-56

[2] Biswas A, Sharma SP. Integrated geophysical studies to elicit the structure associated with uranium mineralization around South Purulia shear zone, India: A review. Ore Geology Reviews. 2016;**72**:1307-1326

[3] Biswas A, Sharma SP. Interpretation of self-potential anomaly over 2-D inclined thick sheet structures and analysis of uncertainty using very fast simulated annealing global optimization. Acta Geodaetica et Geophysica. 2016;**52**(4):439-455

[4] Biswas A, Mandal A, Sharma SP, Mohanty WK. Delineation of subsurface structure using self-potential, gravity and resistivity surveys from South Purulia shear zone, India: Implication to uranium mineralization. Interpretation. 2014;**2**(2):T103-T110

[5] Biswas A. Identification and resolution of ambiguities in interpretation of self-potential data: Analysis and integrated study around South Purulia Shear Zone, India [PhD thesis]. Department of Geology and Geophysics, Indian Institute of Technology Kharagpur; 2013. 199 pp. DOI: http://www.idr.iitkgp.ac.in/xmlui/handle/123456789/3247

[6] Essa K, Mahanee S, Smith PD. A new inversion algorithm for estimating the best fitting parameters of some geometrically simple body to measured self-potential anomalies. Exploration Geophysics. 2008;**39**:155-163

[7] Mehanee S. An efficient regularized inversion approach for self-potential data interpretation of ore exploration using a mix of logarithmic and non-logarithmic model parameters. Ore Geology Reviews. 2014;**57**:87-115

[8] Sundararajan N, SrinivasaRao P, Sunitha V. An analytical method to interpret self-potential anomalies caused by 2D inclined sheets. Geophysics. 1998;**63**:1551-1555

[9] Corry CE. Spontaneous potential associated with porphyry sulphide mineralization. Geophysics. 1985;**50**:1020-1034

[10] Heinrichs WE. Geophysical investigation, ore knob mine, Ashe County, North Carolina. In: Hanson DA, Heinrichs WE, Holmer RC, MacDougall RE, Rogers Summer JS, Ward SH, editors. Mining Geophysics Volume 1, Case Histories. Tusla: SEG; 1966. pp. 179-184

[11] Logn O, Bolviken B. Self-potentials at the Joma pyrite deposit. Geoexploration. 1974; **12**:11-28

[12] Sato M, Mooney HM. The electrochemical mechanism of sulfide self-potentials. Geophysics. 1960;**25**:226-249

[13] Mendonca CA. Forward and inverse self-potential modeling in mineral exploration. Geophysics. 2008;**73**:F33-F43

[14] Monteiro Santos FA, Almeida EP, Castro R, Nolasco M, Mendes-Victor L. A hydrogeological investigation using EM34 and SP surveys. Earth Planets and Space. 2002;**54**:655-662

[15] Rizzo E, Suski B, Revil A, Straface S, Troisi S. Self-potential signals associated with pumping tests experiments. Journal of Geophysical Research. 2004;**109**:B10203

[16] Titov K, Levitski A, Konosavski PK, Tarasov AV, Ilyin YT, Bues MA. Combined application of surface geoelectrical methods for groundwater-flow modeling: A case history. Geophysics. 2005;**70**(5):H21-H31

[17] Jardani A, Revil A, Boleve A, Dupont JP. Three-dimensional inversion of self-potential data used to constrain the pattern of groundwater flow in geothermal fields. Journal of Geophysical Research-Solid Earth. 2008;**113**:B09204

[18] Minsley BJ, Coles DA, Vichabian Y, Morgan FD. Minimization of self-potential survey mis-ties acquired with multiple reference locations. Geophysics. 2008;**73**:F71-F81

[19] Zlotnicki J, Nishida Y. Review on morphological insights of self-potential anomalies on volcanoes. Surveys in Geophysics. 2003;**24**:291-338

[20] Darnet M, Marquis G. Modelling streaming potential (SP) signals induced by water movement in the vadose zone. Journal of Hydrology. 2004;**285**:114-124

[21] Sailhac P, Darnet M, Marquis G. Electrical streaming potential measured at the ground surface: Forward modeling and inversion issues for monitoring infiltration and characterizing the vadose zone. Vadose Zone Journal. 2004;**3**:1200-1206

[22] Hunter LE, Powers MH. Geophysical investigations of earthen dams: An overview. In: 21st EEGS Symposium on the Application of Geophysics to Engineering and Environmental Problems. Poster Overviews II. 2008. pp. 1083-1096

[23] Jardani A, Revil A, Bolève A, Crespy A, Dupont JP, Barrash W. Tomography of the Darcy velocity from self-potential measurements. Geophysical Research Letters. 2007;**34**:1-6

[24] Maineult A. Estimation of the electrical potential distribution along metallic casing from surface self-potential profile. Journal of Applied Geophysics. 2016;**129**:66-78

[25] Drahor MG. Application of the self-potential method to archaeological prospection: Some case histories. Archaeological Prospection. 2004;**11**:77-105

[26] Vichabian Y, Morgan FD. Self-potentials in cave detection. The Leading Edge. 2002; **21**:866-871

[27] Jouniaux L, Pozzi JP. Streaming potential and permeability of saturated sandstones under triaxial stress: Consequences for electrotelluric anomalies prior to earthquakes. Journal of Geophysical Research. 1995;**100**(B6):10197-10209

[28] Kulessa B, Hubbard B, Brown GH. Cross-coupled flow modeling of coincident streaming and electrochemical potentials and application to sub-glacial self-potential data. Journal of Geophysical Research. 2003;**108**(B8):2381

[29] Titov K, Revil A, Konosavsky P, Straface S, Troisi S. Numerical modelling of self-potential signals associated with a pumping test experiment. Geophysical Journal International. 2005;**162**:641-650

[30] Schiavone D, Quarto R. Self-potential prospecting in the study of water movements. Geoexploration. 1984;**22**:47-58

[31] Revil A, Cary L, Fan Q, Finizola A, Trolard F. Self-potential signals associated with preferential ground water flow pathways in a buried paleo-channel. Geophysical Research Letters. 2005;**32**:L07401. DOI: 10.1029/2004GL022124

[32] Di Maio R, Patella D. Self-potential anomaly in volcanic areas: The Mt. Etna case history. Acta Vulcanologica. 1994;**4**:119-124

[33] Di Maio R, Di Sevo V, Gianmmetti S, Patella D, Piscitelli S, Silenziario C. Self-potential anomalies in some Italian volcanic areas. Annals of Geophysics. 1996;**39**:179-188

[34] Mehanee S. Tracing of paleo-shear zones by self-potential data inversion: Case studies from the KTB, Rittsteig, and Grossensees graphite-bearing fault planes. Earth, Planets and Space. 2015;**67**:14

[35] Biswas A, Sharma SP. Interpretation of self-potential anomaly over idealized body and analysis of ambiguity using very fast simulated annealing global optimization. Near Surface Geophysics. 2015;**13**(2):179-195

[36] Biswas A, Sharma SP. Resolution of multiple sheet-type structures in self-potential measurement. Journal of Earth System Science. 2014;**123**(4):809-825

[37] Biswas A, Sharma SP. Optimization of self-potential interpretation of 2-D inclined sheet-type structures based on very fast simulated annealing and analysis of ambiguity. Journal of Applied Geophysics. 2014;**105**:235-247

[38] El-Araby HM. A new method for complete quantitative interpretation of self-potential anomalies. Journal of Applied Geophysics. 2004;**55**:211-224

[39] Eppelbaum L, Khesin B, Itkis S, Ben-Avraham Z. Advanced analysis of self-potential data in ore deposits and archaeological sites. Near surface geoscience. In: 10th European Meeting of Environmental and Engineering Geophysics. Utrecht, The Netherlands; 2004. pp. 1-4

[40] Mehanee S, Essa KS, Smith PD. A rapid technique for estimating the depth and width of a two-dimensional plate from self-potential data. Journal of Geophysics and Engineering. 2011;**8**:447-456

[41] Meiser P. A method of quantitative interpretation of self-potential measurements. Geophysical Prospecting. 1962;**10**:203-218

[42] Murthy BVS, Haricharan P. Nomograms for the complete interpretation of spontaneous potential profiles over sheet like and cylindrical 2D structures. Geophysics. 1985; **50**:1127-1135

[43] Paul MK. Direct interpretation of self-potential anomalies caused by inclined sheets of infinite extension. Geophysics. 1965;**30**:418-423

[44] Pekşen E, Yas T, Kayman YA, Őzkan C. Application of particle swarm optimization on self-potential data. Journal of Applied Geophysics. 2011;**75**:305-318

[45] Rao AD, Babu RHV. Quantitative interpretation of self-potential anomalies due to two-dimensional sheet-like bodies. Geophysics. 1983;**48**:1659-1664

[46] Rao AD, Babu H, Sivakumar Sinha GD. A Fourier transform method for the interpretation of self-potential anomalies due to two-dimensional inclined sheet of finite depth extent. Pure and Applied Geophysics. 1982;**120**:365-374

[47] Roy A, Chowdhury DK. Interpretation of self-potential data for tabular bodies. Journal of Scientific and Engineering Research. 1959;**3**:35-54

[48] Abdelrahman EM, Saber HS, Essa KS, Fouad MA. A least-squares approach to depth determination from numerical horizontal self-potential gradients. Pure and Applied Geophysics. 2004;**161**:399-411

[49] El-Kaliouby HM, Al-Garani MA. Inversion of self-potential anomalies caused by 2D inclined sheets using neural networks. Journal of Geophysics and Engineering. 2009;**6**:29-34

[50] Jagannadha RS, Rama RP, Radhakrishna MIV. Automatic inversion of self-potential anomalies of sheet-like bodies. Computers and Geosciences. 1993;**19**:61-73

[51] Murthy IVR, Sudhakar KS, Rao PR. A new method of interpreting self- potential anomalies of two-dimensional inclined sheets. Computers and Geosciences. 2005;**31**:661-665

[52] Tlas M, Asfahani J. A best-estimate approach for determining self-potential parameters related to simple geometric shaped structures. Pure and Applied Geophysics. 2007;**164**:2313-2328

[53] Tlas M, Asfahani J. An approach for interpretation of self-potential anomalies due to simple geometrical structures using flair function minimization. Pure and Applied Geophysics. 2013;**170**:895-905

[54] Babu HVR, Rao DA. A rapid graphical method for the interpretation of the self-potential anomaly over a two-dimensional inclined sheet of finite depth extent. Geophysics. 1988;**53**:1126-1128

[55] Murthy BVS, Haricharan P. Self-potential anomaly over double line of poles—Interpretation through log curves. Proceedings of the Indian Academy of Science (Earth and Planetary Science). 1984;**93**:437-445

[56] Rao BSR, Murthy IVR, Reddy SJ. Interpretation of self-potential anomalies of some simple geometrical bodies. Pure and Applied Geophysics. 1970;**78**:60-77

[57] Abdelrahman EM, Essa KS, Abo-Ezz ER, Sultan M, Sauck WA, Gharieb AG. New least-square algorithm for model parameters estimation using self- potential anomalies. Computers and Geosciences. 2008;**34**:1569-1576

[58] Roy SVS, Mohan NL. Spectral interpretation of self-potential anomalies of some simple geometric bodies. Pure and Applied Geophysics. 1984;**78**:66-77

[59] Patella D. Introduction to ground surface self-potential tomography. Geophysical Prospecting. 1997;**45**(4):653-681

[60] Patella D. Self-potential global tomography including topographic effects. Geophysical Prospecting. 1997;**45**:843-863

[61] Abdelrahman EM, El-Araby HM, Hassanein AG, Hafez MA. New methods for shape and depth determinations from SP data. Geophysics. 2003;**68**:1202-1210

[62] Abdelrahman EM, El-Araby TM, Essa KS. Shape and depth determinations from second moving average residual self-potential anomalies. Journal of Geophysics and Engineering. 2009;**6**:43-52

[63] Monteiro Santos FA. Inversion of self-potential of idealized bodies anomalies using particle swarm optimization. Computers and Geosciences. 2010;**36**:1185-1190

[64] Fedi M, Abbas MA. A fast interpretation of self-potential data using the depth from extreme points method. Geophysics. 2013;**78**:E107-E116

[65] Li X, Yin M. Application of differential evolution algorithm on self-potential data. PLoS One. 2012;**7**(12):e51199

[66] Di Maio R, Piegari E, Rani P, Avella A. Self-potential data inversion through the integration of spectral analysis and tomographic approaches. Geophysical Journal International. 2016;**206**:1204-1220

[67] Di Maio R, Rani P, Piegari E, Milano L. Self-potential data inversion through a genetic-Price algorithm. Computers and Geosciences. 2016;**94**:86-95

[68] Essa K, El-Hussein M. A new approach for the interpretation of self-potential data by 2-D inclined plate. Journal of Applied Geophysics. 2017;**136**:455-461

[69] Di Maio R, Piegari E, Rani P. Source depth estimation of self-potential anomalies by spectral methods. Journal of Applied Geophysics. 2017;**136**:315-325

[70] Srivastava S, Agarwal BNP. Inversion of the amplitude of the two-dimensional analytic signal of magnetic anomaly by the particle swarm optimization technique. Geophysical Journal International. 2010;**182**:652-662

[71] Nabighian MN. The analytic signal of two-dimensional magnetic bodies with polygonal cross-section, its properties and use for automated anomaly interpretation. Geophysics. 1972;**37**:507-517

[72] Nabighian MN, Grauch VJS, Hansen RO, LaFehr TR, Li Y, Peirce JW, et al. 75th anniversary. The historical development of the magnetic method in exploration. Geophysics. 2005;**70**:33ND-61ND

[73] Nabighian M N, Ander M E, Grauch V JS, Hansen R O, LaFehr T R, Li Y, et al. 75th anniversary. The historical development of the gravity method in exploration. Geophysics. 2005;**70**:63ND-89ND

[74] Nettleton LL. Elementary Gravity and Magnetic for Geologists and Seismologists. Tulsa, OK: SEG; 1971

[75] Srivastava S, Datta D, Agarwal BNP, Mehta S. Applications of ant colony optimization in determination of source parameters from total gradient of potential fields. Near Surface Geophysics. 2014;**12**:373-389

[76] Ingber L, Rosen B. Genetic algorithms and very fast simulated reannealing: A comparison. Mathematical and Computer Modeling. 1992;**16**(11):87-100

[77] Sen MK, Stoffa PL. Global Optimization Methods in Geophysical Inversion. 2nd ed. London: Cambridge Publisher; 2013

[78] Sharma SP, Biswas A. Interpretation of self-potential anomaly over 2D inclined structure using very fast simulated annealing global optimization—An insight about ambiguity. Geophysics. 2013;**78**(3):WB3-W15

[79] Biswas A. Interpretation of gravity and magnetic anomaly over thin sheet-type structure using very fast simulated annealing global optimization technique. Modeling Earth Systems and Environment. 2016;**2**(1):30

[80] Biswas A. A comparative performance of least square method and Very fast simulated annealing global optimization method for interpretation of self-potential anomaly over 2-D inclined sheet type structure. Journal of the Geological Society of India. 2016;**88**(4):493-502

[81] Biswas A, Parija MP, Kumar S. Global nonlinear optimization for the interpretation of source parameters from total gradient of gravity and magnetic anomalies caused by thin dyke. Annals of Geophysics. 2017;**60**(2):G0218, 1-17

[82] Biswas A. Interpretation of residual gravity anomaly caused by a simple shaped body using very fast simulated annealing global optimization. Geoscience Frontiers. 2015;**6**(6):875-893

[83] Dosso SE, Oldenburg DW. Magnetotelluric appraisal using simulated annealing. Geophysical Journal International. 1991;**106**:370-385

[84] Rothman DH. Nonlinear inversion, statistical mechanics and residual statics estimation. Geophysics. 1985;**50**:2784-2796

[85] Rothman DH. Automatic estimation of large residual statics correction. Geophysics. 1986;**51**:337-346

[86] Sharma SP. VFSARES—A very fast simulated annealing FORTRAN program for interpretation of 1-D DC resistivity sounding data from various electrode array. Computers and Geosciences. 2012;**42**:177-188

[87] Biswas A. Inversion of source parameters from magnetic anomalies for mineral /ore deposits exploration using global optimization technique and analysis of uncertainty. Natural Resources Research. 2018;**27**(1):77-107

[88] Biswas A, Acharya T. A Very Fast Simulated Annealing (VFSA) method for inversion of magnetic anomaly over semi-infinite vertical rod-type structure. Modeling Earth Systems and Environment. 2016;**2**(4):198

[89] Mosegaard K, Tarantola A. Monte Carlo sampling of solutions to inverse problems. Journal of Geophysical Research. 1995;**100**(B7):12431-12447

[90] Sen MK, Stoffa PL. Bayesian inference, Gibbs sampler and uncertainty estimation in geophysical inversion. Geophysical Prospecting. 1996;**44**:313-350

[91] Yungul S. Interpretation of spontaneous polarization anomalies caused by spherical ore bodies. Geophysics. 1950;**15**:237-246

[92] Agarwal BNP. Quantitative interpretation of self-potential anomalies. In: Expanded Abstract Volume of the $54^{th}$ SEG Annual Meeting and Exposition; Atlanta. 1984. pp. 154-157

[93] Agarwal B, Srivastava S. Analyses of self-potential anomalies by conventional and extended Euler deconvolution techniques. Computers and Geosciences. 2009;**35**:2231-2238

[94] Bhattacharya BB, Roy N. A note on the use of nomograms for self-potential anomalies. Geophysical Prospecting. 1981;**29**:102-107

[95] Srivastava S, Agarwal BNP. Interpretation of self-potential anomalies by enhanced local wave number technique. Journal of Applied Geophysics. 2009;**68**:259-268

[96] Sundararajan N, Srinivas Y. A modified Hilbert transform and its application to self-potential interpretation. Journal of Applied Geophysics. 1996;**36**:137-143

[97] Telford WM, Geldart LP, Sheriff RE. Applied Geophysics. London: Cambridge University Press; 1990

Section 2

# Importance of Some Economic Mineral Deposits

Chapter 3

# A Review of the Role of Natural Clay Minerals as Effective Adsorbents and an Alternative Source of Minerals

Aref Alshameri, Xinghu Wei, Hailong Wang, Yang Fuguo, Xin Chen, Hongping He, Chunjie Yan and Feng Xu

Additional information is available at the end of the chapter

http://dx.doi.org/10.5772/intechopen.87260

**Abstract**

The minerals with unique properties such as natural clay minerals (NCMs) have promising approach in environmental and industrial sphere. In fact, under some specific conditions the NCMs could be used either as effective adsorbent material or alternative source of minerals. This chapter presents an outline of a general review of factors that affect the application ability of NCMs and a descriptive analysis of $NH_4^+$ and REE adsorption behavior and extraction of rare earth elements (REE) by an ion-exchange with $NH_4^+$ ions onto NCMs. Clays and NCMs both effectively remove various contaminants from aqueous solution and serve as alternative sources of minerals, as extensively discussed in this chapter. This review compiles thorough literature of current research and highlights the key findings of adsorption ($NH_4^+$ and REE) that use different NCMs as adsorbents or alternative sources of minerals (i.e., REE). The review confirmed that NCMs excellently remove different cations pollutants and have significant potential as alternative source of REE. However, modification and further development of NCMs applications for getting the best adsorption and the best extraction of REE onto NCMs, which would enhance pollution control and leaching system is still needed.

**Keywords:** natural clay minerals, adsorption mechanism, ammonium, rare earth elements, REE extraction

# 1. Introduction

Natural clay minerals (NCMs) have gained considerable attention due to their unique properties and their use in huge range of industrial and environmental applications [1, 2]. NCMs are unique in the sense that these minerals are studied by, and used in, many disciplines for essential and applied research [3, 4]. These minerals are nontoxic to ecosystem and play important role in the development of human civilization. They have been utilized in agricultural applications, engineering and construction applications, environmental remediation, geology, pharmaceuticals, food processing and many other industrial applications [2, 5]. The economic benefits look evident due to the fact that NCMs are widespread, and inexpensive compared with other raw materials [6]. For these reasons, NCMs research is being actively pursued by many scientists and in several countries, and the future of clay science seems exciting, and promising.

## 1.1. Structure and composition of the NCMs

NCMs can arguably be considered as phyllosilicate class, containing layered structures of shared octahedral aluminum and tetrahedral silicon sheets, water molecules and hydrated cations that can move in and out of the interlayer spaces [7, 8]. Commonly, isomorphous substitution of one cation with another (of similar size but with lesser charge, such as $Al^{3+}$ for $Si^{4+}$ or $Mg^{2+}$ for $Al^{3+}$) within crystal structures leads to a charge imbalance in silicate NCMs, which accounts for the permanent negative charge on NCMs particles, hence the ability of clays to attract cations to the surface. Amphoteric—OH groups at the surface/edge of clays (i.e., silanol and aluminol groups) could also contribute to surface charge (pH-dependent reversible charge).

The physical and chemical properties of any particular NCMs are structure and composition dependent. A general review of the structure and composition of the various NCMs are essentially hydrous aluminum silicates that sometimes with variable amounts of iron, magnesium, alkali metals, alkaline earths, and other cations found on or near some planetary surfaces [9]. The atomic structure of NCMs consists of two basic units, an octahedral sheet and a tetrahedral sheet. The octahedral sheet is comprised of closely packed oxygens and hydroxyls in which aluminum, iron, and magnesium atoms are arranged in octahedral coordination (**Figure 1a**). The second structural unit is the silica tetrahedral layer in which the silicon atom

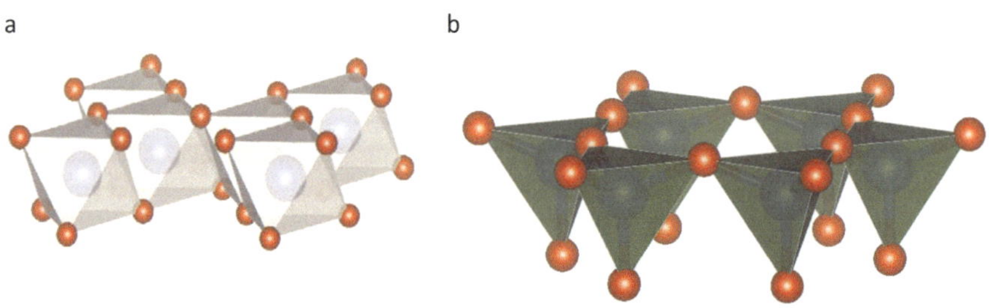


**Figure 1.** Diagrammatic sketch of the octahedral (a) and the tetrahedral sheet (b).

is equidistant from four oxygens or possibly hydroxyls arranged in a tetrahedron with the silicon atom in the center (**Figure 1b**). NCMs are usually classified based on their structure and layer type [10]. The classification of Grim becomes the basis for outlining the nomenclature and the differences between the various NCMs [3]. Although, it is not possible to compress the discussion of structure and classification of NCMs in this chapter, a simple classification of NCMs is available in literatures, NCMs can be divided into four main groups: kaolinite group, illite group, smectite group, and vermiculite group.

The structure and composition of the major industrial NCMs are different even though they are each comprised of octahedral and tetrahedral sheets as their basic building blocks. The arrangement and composition of the octahedral and tetrahedral sheets account for most of the differences in their physical and chemical properties. Therefore, for their applications an understanding of the structure, physical and chemical properties attributes of the individual clay minerals is important. The huge variety of physical and chemical properties of NCMs provides unlimited scope for future application, particularly in environment protection or as minerals resources. A general review of the structure and composition of the NCMs are given in this chapter. A more detailed discussions of the structures of the various NCMs were discussed in literature [5, 9–13].

### 1.2. Properties and factors that affect the application ability of natural clay minerals

The use of NCMs for specific applications depends on its type of structure (1.1 or 2.1 layer type) and on its chemical composition [14]. The identity of all the material present in NCMs should be determined in order to evaluate their properties. The most important characteristic of NCMs is the cation exchange capacity (CEC). CEC is to measure the capacity of NCMs to exchange cations from the solution [15], which depends on the volume of the total layer charge. Since the surface layer charge is the function of pH, thus, CEC also changes with pH and regularly CEC is measured at pH 7 [16, 17]. The popular metallic cations found in exchange positions in NCMs are $Ca^{2+}$, $Mg^{2+}$, $Na^{+}$, and $K^{+}$. The presence of charge in NCMs play important role for cation exchange and the swelling properties of the minerals. The hydrolysis of Si—OH or Al—OH bonds along the NCMs lattices supplies the surface charge. Depending on the silica structure and the solution pH, the net surface charge can be either positive or negative. The tetrahedral and octahedral sheets of NCMs usually have a charge. The charge in the NCMs occurs in two forms: structural and surface charge. The structural charge is permanent and present due to ion substitutions, which arises inside the interior of the layers. The surface charges, generally in NCMs depend on the pH value, with 2:1 layer, the surface charge creates on the basal surface of the tetrahedral sheets while the surface charge for layer type 1:1, derives from both of tetrahedral and octahedral sheets. Also, the edges of both 1:1 and 2:1 layer contribute to the surface charge [16]. Several books have explained the details of the structure and properties of NCMs such as Handbook of Clay Science, edited by Bergaya et al. [13].

Furthermore, the presence of amount of exchangeable ions, non-NCMs, soluble salts, and quality of their texture are factors which can affect their properties and applications. The presence of cations in octahedral sheet, and isomorphic substitutions in the octahedral and tetrahedral sheets result in net charge deficits. Varying according to the sheet unit, and ultimately, in different mineral phases giving rise to varied technical behavior. The textural differences between structurally and chemically identical NCMs also affect their adsorptive properties [15, 17, 18].

The type and amount of non-clay minerals are present with NCMs affect their properties and applications. Non-clay minerals commonly associated with the NCMs include quartz, feldspar, mica, calcite, dolomite, opal C-T, and minor amounts of heavy and trace minerals such as ilmenite, rutile, brookite, anatase, leucoxene, sphene, tourmaline, zircon, kyanite, goethite, hematite, magnetite, garnet, augite, florencite, apatite, andalusite, and barite. Subsequently, when developing applications for NCMs, it is necessary to take these factors into consideration. It is important to know the specific properties of NCMs one is using in order to ensure that it is appropriate for one's needs or to better understand their mechanism behavior during experimental process.

### 1.3. Techniques of NCMs characterization

There is increasing trend in the popularity of productive research in the field of NCMs. Characterization of NCMs is given significance as they are mainly used as cation exchangers, sorbents/hosts, and catalysts. Usually, the characterization of a number of techniques has to be done in order to get comprehensive details of the NCMs.

Moreover, the multitude of techniques also accelerates the process of development of NCMs, particularly as catalysts, as different aspects are discovered. Extensive research in the field of instrumentation has resulted in advanced techniques of analysis that have helped in characterization of molecular sieves in general and NCMs in particular. In common, the characterization of NCMs should provide information about; (i) chemical composition; (ii) structure and morphology; (iii) ability to adsorb and retain molecules, and (iv) ability to chemically convert these molecules.

The techniques for NCMs characterization, include X-ray diffraction (XRD), transmission electron microscopy (TEM), electron-beam-based microscopy, scanning electron microscopy (SEM) with energy-dispersive X-ray spectroscopy (EDX), Fourier transform infrared (FTIR) spectroscopy, X-ray fluorescence (XRF), magic angle spinning (MAS) nuclear magnetic resonance (NMR) spectroscopy (MASNMR), X-ray photoelectron spectroscopy (XPS) analyses, $N_2$ adsorption-desorption isotherms and zeta potential analysis to obtain the mineralogical and physicochemical parameters.

XRD is generally used to identify the presence and quantitative determination of crystalline NCMs. SEM is useful for morphological analysis and chemical analyses at specific locations. The chemical compositions are determined using XRF. IR and NMR provide insight into acid sites and framework structure. It is not possible to compress the discussion of such techniques and research done on their scope and development. Literatures provide useful insight into various characterization techniques for NCMs [5, 13, 15].

Here, we compile research data from various publications related to the use of different NCMs-based adsorbents and the leaching technology generations for REE extraction, notably thorough literature of our current researches. In the first part of this chapter, general structure, properties and the factors that affect the application ability of NCMs are discussed. The techniques of NCMs characterization are also summarized. The main goal of this review is to explain why an understanding of the structure and surface properties of the individual NCMs are so important. This chapter provides an elaborate information about the different NCMs as effective adsorbents in environment protection and their importance for the extraction of rare earth elements in ion adsorption clays. It has also adequately summarized the role of factors that affect adsorption (i.e., $NH_4^+$ and REE) and extraction behavior of REE onto NCMs.

## 2. Adsorption behavior of contaminants and recovery of rare earth elements adsorbed on natural clay minerals

In environment protection, NCMs have been used in the elimination and storage of hazardous chemicals [14, 19, 20]. NCMs have the catalytic capability to neutralize certain organic hazardous chemicals [15, 20, 21]. The ability of NCMs to contain hazardous substances depends on their cation exchange capacity (CEC) while the process of retaining toxic materials mainly occurs by the ion exchange and/or adsorption. Due to their high cation exchange capacity, NCMs are very effective for the adsorption of cations from the solution. Although the NCMs are electronegatively charged [22], these minerals still can adsorb organic and non-ionic substances in significant amounts. The adsorption characteristics are dependent upon the chemical/structural makeup of the adsorbent, the Si/Al ratio, cation type; number and location are particularly influential in adsorption. The adsorption capacity of NCMs can be improved by modification with inorganic salts (NaCl, $CaCl_2$, $BaCl_2$, $NH_4Cl$, AlCl, $FeCl_3$), cationic surfactants, acid, base and organic [23, 24]. Consequently, the NCMs become hydrophobic and organophilic, which led to the enhancement of the adsorption of non-ionic and organic compounds [25, 26]. Notably, in our brief studies of $NH_4^+$ and REE adsorption, we used non-modified NCMs, to better understand the natural system reactivity of NCMs, and also the modification of natural minerals at a larger scale may increase the processing cost; thus, reliable water treatment by using non-modified clay minerals is highly desirable. Therefore, utilization of NCMs would solve disposal problem, and also access to inexpensive materials in the wastewater treatment. Moreover, due to low cost of NCMs, there is no need to regenerate them; which provide more advantages in using NCMs as an adsorbent materials.

### 2.1. A brief review of adsorption of $NH_4^+$

Nitrogen compounds in aqueous environments are commonly found in the form of ammonium ions ($NH_4^+$). Important sources of $NH_4^+$ include effluent from the application in agricultural practices and industrial processes resulting in algal bloom in lakes and rivers [27, 28]. $NH_4^+$ concentration, in certain surface waters serving as a source of potable water, is much higher than the permissible level, due to large quantities of industrial and municipal wastewater being discharged into existing water resources [29–31]. Also, the $NH_4^+$ concentration for most fish species must not exceed 1.5 mg [32, 33]. Therefore, complete removal of $NH_4^+$ is required due to its toxicity to the majority of aquatic lives.

For these reasons, the prevention of nitrogen pollution with $NH_4^+$ removal from wastewater is of great importance [34–36]. Various methods including air stripping, biological methods and activated carbon have been used for $NH_4^+$ removal [37]. However, these techniques are not suitable for use in the removal of low contaminants concentrations, which cause damage both to the environment and life [31]. Additionally, high costs, poor regeneration and uncertainty of outcome are some of the frequently encountered limitations in the application of these methods [36, 38, 39]. Furthermore, contingency on temperature and climate conditions constitutes another disadvantage in this process.

Compared with the above mentioned methods, high safety, low cost [34, 40, 41] and relative simplicity of application and operation are some of the attributes that are attracting an increasing focus on the use of adsorption method for environment applications [33, 42, 43]. Adsorption process is a suitable technique for pollutants removal from wastewater, because of the significant advantages like low-cost, profitability, availability, and effectiveness than other methods. This method is easy to operate and equally effective in the removal of toxic contaminants, even at low concentrations [44].

#### *2.1.1. Investigations of $NH_4^+$ adsorption properties of six natural clay minerals*

This research presented six types of NCMs-based adsorbents namely kaolinite, halloysite, montmorillonite, vermiculite, palygorskite, and sepiolite were examined and compared in the same study [44]. The study illustrated that among all the NCMs studied, vermiculite and montmorillonite have the highest ammonium adsorption capacities. The study revealed that the cation exchange is the main mechanism for the $NH_4^+$ adsorption. Negatively charged surface, specific surface area, water absorption process and surface morphology of NCMs might also contribute to the high adsorption capacities. Adsorption kinetics showed that the adsorption behavior followed the pseudo-second-order kinetic model whereas the isotherms fitted the Langmuir model. The insights obtained in this study are useful for applications of NCMs in environmental remediation. The results illustrated that the structure and surface properties of NCMs are the key factors that affect the adsorption capacities for $NH_4^+$. The study concluded that the NCMs have significant potential as economic, safe and effective adsorbent materials for the $NH_4^+$ adsorption from the aqueous solution at low concentrations.

### 2.2. A brief review of adsorption/extraction of rare earth elements

#### *2.2.1. Adsorption of rare earth elements (REE)*

REE group consists of 17 elements and is divided into two categories namely the light rare earths (L-REE) and the heavy rare earths (H-REE). REE have been used widely in metallurgy, chemical engineering, electronics and electrooptics, medicine, biomedicine, for manufacturing of magnetostriction materials and lasers [45–47]. Its applications in advanced technologies are increasing [48]. In modern societies, the rare earth elements (REE) are considered because of their unique physical and chemical properties. REE will be of substantial attention for the foreseeable future, with demand likely to grow.

http://dx.doi.org/10.5772/intechopen.87260

Therefore, due to the increasing use of REE in industries, determination of REE has been of a recent increasing concern. Several analytical techniques were used to determine REE in samples such as inductively coupled plasma-mass spectrometry (ICP-MS) [49], neutron activation (INAA) [50]. Energy dispersive X-ray fluorescence (EDXRF) [51] and inductively coupled plasma optical emission spectroscopy (ICP-OES) [52]. Research should continue to play an important role in the search for rare earth ore deposits and their extraction, ensuring that as little damage is done to the environment as conceivable.

Various techniques have been used for removal of REE ions from aqueous systems such as solid-phase extraction, solvent extraction, ion exchange, ion-selective electrodes [53–59] and adsorption [60–62]. Adsorption method is the best technique because of low cost, simplicity of design and operation. The ion-adsorption type rare earths ore, is mainly located in China and REE in these deposits were released by weathering of REE-rich granites and subsequently adsorbed by NCMs.

Recently, there has been a significant interest in adsorption of rare earth elements with NCMs [1, 63–65]. Piasecki and Sverjensky [66] also had studied REE speciation/distribution on wide ranges of pH and ionic strength. They concluded that most of the surface-adsorbed lanthanides occur as simple "clay-REE" or as hydrolyzed "clay-O-$REE^{2+}$" species.

NCMs are electronegative, saturated with cations such as $Na^+$, $K^+$, $Ca^{2+}$, and $Mg^{2+}$ are capable of exchanging cations such as REE to the surface. Previous researches have shown that REE contained in NCMs are mainly present as physisorbed ions, which can be easily recovered by a simple ion-exchange procedure [66, 67]. It is evident that the adsorption of REE ions on NCMs would have great influence on the mineralization process and the leaching process of the ion-adsorption type rare earths ore.

### 2.2.2. *Extraction of REE from ion adsorption clays*

The ion-adsorption type rare earths ore are generally formed by weathering of REE rich host rocks (granitic or igneous) and transfer into an aqueous solution which percolates through the weathering body and are adsorbed onto NCMs [68–71]. This provides evidence that NCMs have the ability of adsorbing lanthanide ions released/solubilized during weathering [72]. However, although, the NCMs deposits containing adsorbed lanthanides which are of substantially lower grade than other types of REE mineral resources, the economic benefits look remarkable, because NCMs are abundant in surface layers in nature, easier ionic exchange of REE, ease of mining and processing [66, 73]. It is thus evident that the adsorption of rare earth ions on NCMs would have great influence on the mineralization process and the leaching process of the ion-adsorption type rare earths ore.

During the past 45 years, the leaching technology in REE extraction were investigated [64, 74, 75]. The extraction of REE has been a long tradition in China using the concept of ion-exchange leaching. In the ion adsorption clays 60–90% of the REE are adsorbed onto NCMs [71] and the adsorbed REE on NCMs could be easily recovered by leaching with monovalent salt solutions. Recently, the ion-adsorption rare earth ores have the focus of most research endeavors as an alternative source of REE, the results of leaching efficiencies has been reported [66, 71, 73, 75, 76]. Based on these findings, $(NH_4)_2SO_4$ was identified as the best a lixiviant of REE from NCMs.

*2.2.3. Investigations of adsorption/extraction behavior for REE onto natural clay minerals*

Four natural clay minerals namely kaolinite, montmorillonite, muscovite and illite were systematically investigated and compared for their adsorption/extraction behavior for REE [77]. The study reported that the montmorillonite exhibits highest adsorption and regeneration efficiencies for REE while kaolinite has highest extractions efficiencies for both REE light and heavy in the order of kaolinite > illite > montmorillonite > muscovite. Also study found that the lack extractions of REE from muscovite than other NCMs are believed to presence of iron oxide and biotite mineral (produce iron oxide as a result of its alteration) associated with muscovite. The leaching process of the REE is a kind of the reversibility of the ion-exchange process, it was evident that the cation exchange and negatively charged surfaces are the mechanism for REE adsorption. It was concluded that $NH_4^+$ is lacking as a lixiviant from NCMs since NCMs are associated with iron oxide, particularly, either with NCMs containing iron (i.e., biotite) or minerals which always associate with biotite such as muscovite. The important role of the pH in extraction of REE from NCMs was evidenced, when REE-NCMs come into contact with the $NH_4^+$ solution, the pH is rapidly increased from the initial pH solution for both montmorillonite and muscovite, leading to the decrease of the availability of ion-exchangeable REE with $NH_4^+$ ions. That is one of the factors that influence the reduction in the REE extraction from montmorillonite and muscovite when compared with those of kaolinite and illite. The results illustrated that the structure and surface properties of NCMs are also the key factors that affect the rare earth leaching, consequently identifying the types of NCMs and associated impurities in clay materials is important for getting the best leaching system [77].

## 3. Conclusion

NCMs have gained a significant interest among the scientific community, because of their abundance, low cost and their unique properties. In this regard, a systematic comparison study under identical experimental conditions could help to better understand the influence of structure and properties of NCMs on their adsorptive/extraction behaviors towards contaminants and elements in leaching process. Adsorption is a very promising and efficient technology for the removal of hazardous contaminants from water source, thus NCMs have been successfully used as an adsorbent materials and alternative source of minerals. Batch adsorption experiments have demonstrated that the contact time, initial pollutants concentration, adsorbent dosage and solution pH have significant effects on contaminants adsorption/desorption. Among all the NCMs studied, montmorillonite and vermiculite exhibit the highest adsorption efficiencies towards $NH_4^+$ and REE. The presence of iron oxide with NCMs found to help enhance REE adsorption (REE-Fe-oxides), meanwhile it also influences the extraction of REE because REE-Fe-oxides cannot be easily recovered by monovalent salt solutions using the concept of ion exchange i.e., $NH_4^+$. The structure and surface properties of NCMs are the key factors that affect the adsorption capacities for contaminants and extraction of minerals (i.e., REE). The review suggests that NCMs can be considered as ideal adsorbents and alternative source of minerals owing to their low cost, abundant, high safety, and good adsorption efficiencies. Identifying the types of NCMs and associated impurities in clay materials is important either for getting the best adsorption or the best leaching system. Thus, when developing applications for NCMs, it is essentially to take these factors into consideration. However, further study is necessary to establish the process parameters to generate better quality of products. Also, modification and thermal treatment of natural clay minerals could provide potential future applications in water treatment.

## Acknowledgements

This work was supported by the scientific research start up fund (099/CGZ07273), School of Environmental and Chemical Engineering, Foshan University, Foshan, Guangdong 528000, China.

## Author details

Aref Alshameri[1,2,3,4]*, Xinghu Wei[1]*, Hailong Wang[1], Yang Fuguo[1], Xin Chen[1], Hongping He[2], Chunjie Yan[4] and Feng Xu[1]

*Address all correspondence to: aref_alshmiri@yahoo.com, weixing1964@163.com

1 School of Environmental and Chemical Engineering, Foshan University, Foshan, Guangdong, China

2 Guangdong Provincial Key Laboratory of Mineral Physics and Materials and Key Laboratory of Mineralogy and Metallogeny, Guangzhou Institute of Geochemistry, Chinese Academy of Sciences, Guangzhou, China

3 Geological Survey and Mineral Resources Board, Ministry of Oil and Minerals, Sana'a, Yemen

4 Engineering Research Center of Nano-Geomaterial of Education Ministry, China University of Geosciences, Wuhan, China

## References

[1] Anastopoulos I, Bhatnagar A, Lima E. Adsorption of rare earth metals: A review of recent literature. Journal of Molecular Liquids. 2016;**221**:954-962. DOI: 10.1016/j.molliq.2016.06.076

[2] Reyes C, Fiallo L. Application of illite-and kaolinite-rich clays in the synthesis of zeolites for wastewater treatment. In: Imran Ahmad Dar (ed.), Earth and Environmental Sciences. Rijeka, Croatia: Intech Open; 2011. p. 234-246. DOI: 10.5772/25895

[3] Grim R. Applied Clay Mineralogy. New York, USA: McGraw-Hill Book Company; 1962. 6d. DOI: 10.1080/11035896209447314

[4] Su L, Zeng X, He H, Tao Q, Komarneni S. Preparation of functionalized kaolinite/epoxy resin nanocomposites with enhanced thermal properties. Applied Clay Science. 2017;**148**: 103-108. DOI: 10.1016/j.clay.2017.08.017

[5] Murray H. Applied Clay Mineralogy. Developments in Clay Science. 1st ed. Amsterdam: Elsevier; 2006. 188p. DOI: 10.1016/S1572-4352(06)02001-0

[6] Sabino D, Giusy L, Mariangela G, Michele N. Characteristics and adsorption capacities of low-cost sorbents for wastewater treatment: A review. Sustainable Materials and Technologies. 2016;**9**:10-40. DOI: 10.1016/j.susmat.2016.06.002

[7] Sposito G, Skipper N, Sutton R, Park S, Soper A, Greathouse J. Surface geochemistry of the clay minerals. Proceedings of the National Academy of Sciences. 1999;**96**:3358-3364. DOI: 10.2307/47645

[8] Bergaya F, Lagaly G. Hand Book of Clay Science, General Introduction: Clays, Clay Minerals, and Clay Science. Developments in Clay Science. Amsterdam: Elsevier; 2013. p.19. DOI: 10.1016/B978-0-08-098258-8.00001-8

[9] Bergaya F, Lagaly G, editors. Handbook of Clay Science. Developments in Clay Science. Amsterdam: Elsevier; 2006. 1224 pp

[10] Brigatti F, Galan E, Theng B. Structures and mineralogy of clay minerals. In: Developments in Clay Science. Amsterdam: Elsevier; 2006. pp. 19-86. DOI: 10.1016/S1572-4352(05) 01002-0

[11] Bailey S. Chlorites; structures and crystal chemistry. Reviews in Mineralogy and Geochemistry. 1988;**19**:347-403. DOI: 10.1515/9781501508998-015

[12] Moore D, Reynolds R. X-Ray Diffraction and the Identification and Analysis of Clay Minerals. Oxford: Oxford University Press; 1989. 322 pp. DOI: 10.1180/claymin.1990.025.4.11

[13] Bergaya F, Lagaly G. General introduction: Clays, clay minerals, and clay science. Developments in Clay Science. 2006;**1**:1-18. DOI: 10.1016/S1572-4352(05)01001-9

[14] Zadinelo I, José Alves H, Moesch A, Colpini L, Silva L, Santos L. Influence of the chemical composition of smectites on the removal of ammonium ions from aquaculture effluents. Journal of Materials Science. 2015;**50**:1865-1875. DOI: 10.1007/s10853-014-8749-3

[15] Ismadji S, Soetaredjo F, Ayucitra A. Clay Materials for Environmental Remediation. Cham: Springer; 2015. DOI: 10.1007/978-3-319-16712-1

[16] Eslinger E, Pevear D. Clay Minerals for Petroleum Geologists and Engineers. SEPM Short Course Notes no 22, ix + 405 pp. Society of Economic Paleontologists and Mineralogists. Tulsa; 1988. OI: https://doi.org/10.2110/scn.88.01

[17] Yuan G, Theng B, Churchman J, Gates W. Clays and clay minerals for pollution control. In: Developments in Clay Science. Amsterdam: Elsevier; 2013. p. 587-644. DOI: 10.1016/B978-0-08- 098259-5.00021-4

[18] Xinggang W, Shaoyu L, Chunmei G, Xiubing X, Xinjie Z, Xiao B, et al. Highly efficient adsorption of ammonium onto palygorskite nanocomposite and evaluation of its recovery as a multifunctional slow-release fertilizer. Chemical Engineering Journal. 2014;**252**:404-414. DOI: 10.1016/j.cej.2014.04.097

[19] Lin S, Juang R. Adsorption of phenol and its derivatives from water using synthetic resins and low-cost natural adsorbents: A review. Journal of Environmental Management. 2009; **90**:1336-1349. DOI: 10.1016/j.jenvman.2008.09.003

[20] Beall G. The use of organo-clays in water treatment. Applied Clay Science. 2003;**24**:11-20. DOI: 10.1016/j.clay.2003.07.006

[21] Uğurlu M, Karaoğlu M. Adsorption of ammonium from an aqueous solution by fly ash and sepiolite: Isotherm, kinetic and thermodynamic analysis. Microporous and Mesoporous Materials. 2011;**139**:173-178. DOI: 10.1016/j.micromeso.2010.10.039

[22] Churchman G, Gates W, Theng B, Yuan G. Chapter 11.1 Clays and clay minerals for pollution control. In: Developments in Clay Science. Vol. 1. Amsterdam: Elsevier; 2006. p. 625-675. DOI: 10.1016/S15724352(05)01022

[23] Seung L, Diwakar T. Organo and inorgano-organo-modified clays in the remediation of aqueous solutions: An overview. Applied Clay Science. 2012;**59-60**:84-102. DOI: 10.1016/j.clay.2012.02.006

[24] Vercellone S, Sham E, Torres E. Measure of zeta potential of titanium pillared clays. Procedia Materials Science. 2015;**8**:599-607. DOI: 10.1016/j.mspro.2015.04.114

[25] Srinivasan R. Advances in application of natural clay and its composites in removal of biological, organic, and inorganic contaminants from drinking water. Advances in Materials Science and Engineering. 2011;**2011**:1-17. DOI: 10.1155/2011/872531

[26] Zaghouane-Boudiaf H, Boutahala M. Kinetic analysis of 2,4,5-trichlorophenol adsorption onto acid-activated montmorillonite from aqueous solution. International Journal of Mineral Processing. 2011;**100**:72-78. DOI: 10.1016/j.minpro.2011.04.011

[27] Wang M, Liao L, Zhang X, Li Z, Zhiguo X, Cao W. Adsorption of low-concentration ammonium onto vermiculite from Hebei Province, China. Clays and Clay Minerals. 2011;**59**:459-465. DOI: 10.1346/CCMN.2011.0590503

[28] Peng X, Xuejing W, Xin W, Jingke S, Hao W, Jing Z, et al. Struvite crystallization combined adsorption of phosphate and ammonium from aqueous solutions by mesoporous MgO–loaded diatomite. Colloids and Surfaces A: Physicochemical and Engineering Aspects. 2016;**506**:220-227. DOI: 10.1016/j.colsurfa.2016.05.101

[29] Njoroge B, Mwamachi S. Ammonia removal from an aqueous solution by the use of a natural zeoli. Journal of Environmental Engineering and Science. 2004;**3**:147-154. DOI: 10.1139/s03-070

[30] Yanhui L, Qiuju D, Tonghao L, Xianjia P, Junjie W, Jiankun S, et al. Comparative study of methylene blue dye adsorption onto activated carbon, graphene oxide, and carbon nanotubes. Chemical Engineering Research and Design: Transactions of the Institution of Chemical Engineers. 2013;**91**:361-368. DOI: 10.1016/j.cherd.2012.07.007

[31] Alshameri A, Yan C, Al-Ani Y, Dawood A, Ibrahim A, Zhou C, et al. An investigation into the adsorption removal of ammonium by salt activated Chinese (Hulaodu) natural zeolite: Kinetics, isotherms, and thermodynamics. Journal of the Taiwan Institute of Chemical Engineers. 2014;**45**:554-564. DOI: 10.1016/j.jtice.2013.05.008

[32] Sun Z, Qu X, Wang G, Zheng S, Frost R. Removal characteristics of ammonium nitrogen from wastewater by modified Ca-bentonites. Applied Clay Science. 2015;**107**:46-51. DOI: 10.1016/j.clay.2015.02.003

[33] Sarioglu M. Removal of ammonium from municipal wastewater using natural Turkish (Dogantepe) zeolite. Separation and Purification Technology. 2005;**41**:1-11. DOI: 10.1016/j.seppur.2004.03.008

[34] Gupta V, Agarwal S, Saleh T. Synthesis and characterization of alumina-coated carbon nanotubes and their application for lead removal. Journal of Hazardous Materials. 2011; **185**:17-23. DOI: 10.1016/j.jhazmat.2010.08.053

[35] Leyva-Ramos R, Monsivais-Rocha J, Aragon-Piña A, Berber-Mendoza M, Guerrero-Coronado R, Alonso-Davila P, et al. Removal of ammonium from aqueous solution by ion exchange on natural and modified chabazite. Journal of Environmental Management. 2010;**91**:2662-2668. DOI: 10.1016/j.jenvman.2010.07.035

[36] Alshameri A, Ibrahim A, Assabri A, Lei X, Wang H, Yan C. The investigation into the ammonium removal performance of Yemeni natural zeolite: Modification, ion exchange mechanism, and thermodynamics. Powder Technology. 2014;**258**:20-31. DOI: 10.1016/j.powtec.2014.02.063

[37] Moradi O, Yari M, Zare K, Marza B, Najafi F. Carbon nanotubes: A review of chemistry principles and reactions. Fullerene Science and Technology. 2012;**20**:138-151. DOI: 10.1080/1536383X.2010.533312

[38] Europea C. IPPC Reference Document on Best Available Techniques in Common Waste Water and Waste Gas Treatment/Management Systems in the Chemical Sector. Sevilla, Espanya: European IPPC Bureau; 2003

[39] Moussavi G, Talebi S, Farrokhi M, Sabouti R. The investigation of mechanism, kinetic and isotherm of ammonia and humic acid co-adsorption onto natural zeolite. Chemical Engineering Journal. 2011;**171**:1159-1169. DOI: 10.1016/j.cej.2011.05.016

[40] Dimirkou A, Doula M. Use of clinoptilolite and an Fe-over exchanged clinoptilolite in $Zn^{2+}$ and $Mn^{2+}$ removal from drinking water. Desalination. 2008;**224**:280-292. DOI: 10.1016/j.desal.2007.06.010

[41] Alshameri A, Abood A, Yan C, Muhammad AM. Characteristics, modification and environmental application of Yemen's natural bentonite. Arabian Journal of Geosciences. 2014;**7**:841-853. DOI: 10.1007/s12517-013-0855-z

[42] Du Q, Liu S, Cao Z, Wang Y. Ammonia removal from aqueous solution using natural Chinese clinoptilolite. Separation and Purification Technology. 2005;**44**:229-234. DOI: 10.1016/j.seppur.2004.04.011

[43] Kurniawan A, Chan G, Lo W, Babel S. Comparisons of low-cost adsorbents for treating wastewaters laden with heavy metals. Science of the Total Environment. 2006;**366**:409-426. DOI: 10.1016/j.scitotenv.2005.10.001

[44] Alshameri A, He H, Zhu J, Xi Y, Zhu R, Ma L, et al. Adsorption of ammonium by different natural clay minerals: Characterization, kinetics and adsorption isotherms. Applied Clay Science. 2018;**159**:83-93. DOI: 10.1016/j.clay.2017.11.007

[45] Greaves M, Elderfield H, Sholkovitz E. Aeolian sources of rare earth elements to the Western Pacific Ocean. Marine Chemistry. 1999;**68**:31-38. DOI: 10.1016/S0304-4203(01)00018-4

[46] Kondo K, Kamio E. Separation of rare earth metals with a polymeric microcapsule membrane. Desalination. 2002;**144**:249-254. DOI: 10.1016/S0011-9164(02)00320-X

[47] Tadjarodi A, Jalalat V, Zare-Dorabei R. Adsorption of La(III) in aqueous systems by N-(2-hydroxyethyl) salicylaldimine-functionalized mesoporous silica. Materials Research Bulletin. 2015;**61**:113-119. DOI: 10.1016/j.materresbull.2014.09.036

[48] Binnemans K, Jones P. Rare earths and the balance problem. Journal of Sustainable Metallurgy. 2015;**1**:29-38. DOI: 10.1007/s40831-014-0005-1

[49] Raut N, Huang L, Aggarwal S, Lin K. Determination of lanthanides in rock samples by inductively coupled plasma mass spectrometry using thorium as oxide and hydroxide correction standard. Spectrochemical Acta Part B Atomic Spectroscopy. 2003;**58**:809-822. DOI: 10.1016/S0584-8547(03)00016-8

[50] Wang Y, Sun J, Chen H, Guo F. Determination of the contents and distribution characteristics of REE in natural plants by NAA. Journal of Radioanalytical and Nuclear Chemistry. 1997;**219**:99-103. DOI: 10.1007/BF02040273

[51] Cornejo-Ponce L, Peralta-Zamora P, Bueno M. Pre-concentration of rare earths using silica gel loaded with 1-(2-pyridylazo)-2-naphthol (PAN) and determination by energy dispersive X-ray fluorescence. Talanta. 1998;**46**(6):1371-1378. DOI: 10.1016/S0039-9140(98)00002-2

[52] Premadas A, Srivastava P. Inductively coupled plasma atomic emission spectrometric determination of lanthanides and Y in various uranium hydrometallurgical products. Journal of Radioanalytical and Nuclear Chemistry. 2002;**251**:233-239. DOI: 10.1023/A:1014804007992

[53] Ganjali M, Daftari A, Rezapour M, Puorsaberi T, Haghgoo S. Gliclazide as novel carrier in construction of PVC-based La(III)-selective membrane sensor. Talanta. 2003;**59**:613-619. DOI: 10.1016/S0039-9140(02)00573-8

[54] Gupta V, Jain S, Chandra S. Chemical sensor for lanthanum(III) determination using aza-crown as ionophore in poly(vinyl chloride) matrix. Analytica Chimica Acta. 2003;**486**:199-207. DOI: 10.1016/S0003-2670(03)00506-3

[55] Ganjali M, Kiani-Anbouhi R, Shamsipur M, Poursaberi T, Salavati-Niasari M, Talebpour Z, et al. Novel potentiometric PVC-membrane and coated graphite sensors for lanthanum (III). Electroanalysis. 2010;**16**:1002-1008. DOI: 10.1002/elan.200302899

[56] Zhou S, Li X, Shi Y, Alshameri A, Yan C. Preparation, characterization, and Ce(III) adsorption performance of poly (allylamine)/silica composite. Desalination and Water Treatment. 2015;**5**:1321-1334. DOI: 10.1080/19443994.2014.944221

[57] Akhond M, Najafi M, Tashkhourian J. Lanthanum-selective membrane electrode based on 2,2′-dithiodipyridine. Analytica Chemical Acta. 2005;**531**:179-184. DOI: 10.1016/j.aca.2004.09.089

[58] Shamsipur M, Ershad S, Naser S, Esmaeilbeig A, Kia R, Abdolmaleki A. Polymeric membrane lanthanum(III)-selective electrode based on N,N′-adipylbis (5-phenylazo salicylaldehyde hydrazone). Electroanalysis. 2010;**17**:1828-1834. DOI: 10.1002/elan.200503309

[59] Berijani S, Ganjali M, Sereshti H, Norouzi P. A selective modified nanoporous silica as sorbent for separation and preconcentration of dysprosium in water samples prior to ICP-OES determination. International Journal of Environmental Analytical Chemistry. 2012; **92**:355-365. DOI: 10.1080/03067319.2010.551120

[60] Kikuchi T, Nogami M, Suzuki K. Separation performance for trivalent actinides from lanthanides by thiacalix 4 arenes compounds impregnated silica ion-exchanger. Journal of Alloys and Compounds. 2004;**374**:272-276. DOI: 10.1016/j.jallcom.2003.11.101

[61] Perez-Quintanilla D, Del Hierro I, Fajardo M, Sierra I. Cr(VI) adsorption on functionalized amorphous and mesoporous silica from aqueous and non-aqueous media. Materials Research Bulletin. 2007;**42**:1518-1530. DOI: 10.1016/j.materresbull.2006.10.029

[62] Zheng H, Han L, Ma H, Zheng Y, Zhang H, Liu D, et al. Adsorption characteristics of ammonium ion by zeolite 13X. Journal of Hazardous Materials. 2008;**158**:577-584. DOI: 10.1016/j.jhazmat.2008.01.115

[63] Bradbury M, Baeyens B. Sorption of Eu on Na- and Ca-montmorillonites: Experimental investigations and modelling with cation exchange and surface complexation. Geochimica et Cosmochimica Acta. 2002;**66**:2325-2334. DOI: 10.1016/S0016-7037(02)00841-4

[64] Coppin F, Berger G, Bauer A, Castet S, Loubet M. Sorption of lanthanides on smectite and kaolinite. Chemical Geology. 2002;**182**:57-68. DOI: 10.1016/S0009-2541(01)00283-2

[65] Gladysz-Plaska A, Majdan M, Grabias E. Adsorption of La, Eu and Lu on raw and modified red clay. Journal of Radioanalytical and Nuclear Chemistry. 2014;**301**:33-40. DOI: 10.1007/s10967-014-3111-4

[66] Moldoveanu G, Papangelakis V. Recovery of rare earth elements adsorbed on clay minerals: I. Desorption mechanism. Hydrometallurgy. 2012;**117**:71-78. DOI: 10.1016/j.hydromet.2012.02.007

[67] Moldoveanu G, Papangelakis V. Recovery of rare earth elements adsorbed on clay minerals: II. Leaching with ammonium sulfate. Hydrometallurgy. 2013;**131**:158-166. DOI: 10.1016/j.hydromet.2012.10.011

[68] Shaomei P. The geological characteristics and the prospecting criteria of the granite-weathering crust ion adsorption type Ree deposits in Nanling area, South China. Materials Science Forum. Trans Tech Publication. 1991;**70-72**:33-42. DOI: 10.4028/www.scientific.net/MSF.70-72.33

[69] Aja S. The sorption of the rare earth element, Nd, onto kaolinite at 25°C. Clays and Clay Minerals. 1998;**46**:103-109. DOI: 10.1346/CCMN.1998.0460112

[70] Bao Z, Zhao Z. Geochemistry of mineralization with exchangeable REY in the weathering crusts of granitic rocks in South China. Ore Geology Reviews. 2008;**33**:519-535. DOI: 10.1016/j.oregeorev.2007.03.005

[71] Feng J, Zhou F, Chi R, Liu X, Xu Y, Xu Y. Effect of a novel compound on leaching process of weathered crust elution-deposited rare earth ore. Minerals Engineering. 2018;**129**:63-70. DOI: 10.1016/j.mineng.2018.09.018

[72] Bruque S, Mozas T, Rodriguez A. Factors influencing retention of lanthanide ions by montmorillonite. Clay Minerals. 1980;**15**:413-420. DOI: 10.1180/claymin.1980.015.4.08

[73] Papangelakis V, Moldoveanu G. Recovery of rare earth elements from clay minerals. In: 1st European Rare Earth Resource Conference; 04–07 September 2014; Milos. 2014. pp. 191-202

[74] Sinitsyn V, Aja S, AKulik D, AWood S. Acid-base surface chemistry and sorption of some lanthanides on $K^+$-saturated marblehead illite: I. Results of an experimental investigation. Geochemical et Cosmochimica Acta. 2000;**64**:185-194. DOI: 10.1016/S0016-7037(99)00175-1

[75] Chi R, Zhou Z, Xu Z, Hu Y, Zhu G, Xu. Solution-chemistry analysis of ammonium bicarbonate consumption in rare-earth-element precipitation. Metallurgical and Materials Transactions B: Process Metallurgy and Materials Processing Science. 2003;**34**:611-617. DOI: 10.1007/s11663-003-0031-z

[76] Tian J, Yin J, Chen K, Rao G, Jiang M, Chi R. Extraction of rare earths from the leach liquor of the weathered crust elution-deposited rare earth ore with non-precipitation. International Journal of Mineral Processing. 2011;**98**:125-131. DOI: 10.1016/j.minpro.2010.11.007

[77] Alshameri A, He H, Xin C, Zhu J, Xinghu W, Zhu R, et al. Understanding the role of natural clay minerals as effective adsorbents and alternative source of rare earth elements: Adsorption operative parameters. Hydrometallurgy. 2019;**185**:149-161. DOI: 10.1016/j.hydromet.2019.02.016

Chapter 4

# Theoretical Studies of the Structural, Mechanical and Raman Spectroscopic Properties of Uranyl-Containing Minerals

Francisco Colmenero Ruiz

Additional information is available at the end of the chapter

http://dx.doi.org/10.5772/

**Abstract**

The incipient use of theoretical methods in the research of geomaterials reveals the great power of such methodology in the determination of the mineral properties. These methods provide a safe, accurate and cheap manner of obtaining these properties. Uranium-containing minerals are highly radiotoxic, and their experimental studies demand a careful handling of the samples used. However, theoretical methods are free of such inconveniences and may be used in the complete characterization of this type of minerals. Theoretical methods are not only a complement to the use of other experimental techniques but also a powerful predictive tool. The structural, mechanical and Raman spectroscopic properties of uranyl-containing materials, including rutherfordine soddyite, schoepite and uranophane-$\alpha$, were studied by means of theoretical solid-state methods based on density functional theory using plane waves and pseudopotentials. A new norm-conserving relativistic pseudopotential for uranium atom developed in recent works was employed. These minerals are among the most important secondary phases arising from corrosion of spent nuclear fuel under the final geological disposal conditions. The computed crystal structures of these materials as well as the corresponding and X-ray powder patterns were found to be in excellent agreement with the experimental information. Therefore, the optimized structures of these minerals were employed to study the mechanical properties and stability of these minerals. These properties were obtained using the finite deformation technique. All these minerals were found to be mechanically stable since the corresponding Born stability conditions were satisfied. A large amount of relevant mechanical data were reported including bulk, Young and Shear moduli, Poisson ratios, ductility and hardness indices, anisotropy measures as well as longitudinal and transversal wave velocities. The large volume expansion and mechanical stress resulting from the corrosion of spent nuclear fuel during storage emphasize the great relevance of the mechanical information of the waste components. Finally, the computation of vibrational properties of these minerals is studied. The computed Raman spectra of these materials were found to be in good agreement with their experimental counterparts when they were available for comparison. These results demonstrate the power of the theoretical methods in the research of uranium-containing minerals.

**Keywords:** uranyl-containing minerals, spent nuclear fuel, density functional theory, crystal structures, X-ray diffraction, mechanical properties, Raman spectroscopy, rutherfordine, soddyite, uranophane-α, schoepite

## 1. Introduction

Nuclear energy covers a fundamental fraction of the increasing electricity requirements due to the growing populations and consumption. It provides a continuous and sustainable energy source independent of the climate and local conditions and free of $CO_2$ emissions and other greenhouse gases causing global warming [1–3]. However, nuclear power installations produce high-level radioactive nuclear waste (HLRW) whose management is very complicated and can be a source of high ecological damage [1–4]. Special care must be dedicated to the avoidance of nuclear accidents [5]. The need of extreme care to avoid environmental issues must be extended to the uranium ore mining and nuclear fuel production [6–8].

Spent nuclear fuel (SNF) will be stored in an underground deep geological repository (DGR). We expect that the barriers protecting the HLRW will be ineffective after a time period of the order of thousands of years [9] after closure. At this moment, groundwater will enter in contact with the waste and the reducing conditions in the DGR will not be preserved. An oxidative environment will appear in a layer near the SNF surface having a thickness of about 50 μm [10]. The concentration of oxidized species as hydrogen peroxide [11] near the SNF surface will augment because of the radiolysis of water originated by the intense alpha radiation produced by the SNF [12–15]. Uranium in the matrix of the SNF then oxidizes from U(IV) to U(VI) and dissolves into the water forming uranyl groups ($UO_2^{2+}$). Secondary phases, that is, alteration products, on the spent fuel surface will appear due to the precipitation of these uranyl groups. The composition of these secondary phases will depend on the local physico-chemical conditions (mainly the pH and electrochemical potential) and the concentrations of reactive species present [16].

The secondary phases of SNF are more easily studied by analyzing natural uranyl-containing minerals [17] found as alteration products of uraninite [18], since this mineral is a natural analogue of the SNF matrix. The different alteration products of natural uraninites and its paragenetic sequence under different geochemical conditions were first described by Frondel [19, 20], and it is still widely accepted nowadays [16, 18, 21–24]. In this sequence, the uranyl oxide hydrates appear first, then the uranyl silicates and, less frequently, the uranyl phosphates, although the specific alteration products depend on the local conditions. Uranyl carbonates may precipitate where the evaporation is significant, and the carbon dioxide partial pressure is large [17, 25].

The release and the concomitant environmental impact of the fission products and transuranic elements present in the SNF to the biosphere [8, 26–41] can be diminished by retention processes of these contaminants in the crystal structures of these secondary phases. The precise knowledge of their unit cells [27, 28] is essential because it may be a used to understand and evaluate the incorporation of these elements into their crystal structures [8, 26–41]. Besides, the long-term performance assessment of DGRs requires the development of identification procedures and the characterization of the physical properties of these alteration products, which is a great challenge from the experimental point of view not only because these materials are very complicated involving the most elements of the periodic table but also due to its radiotoxicity [1].

An experimental technique is appropriate for the identification of uranyl-containing materials if fulfills the following demands [42]: (1) the samples do not require any special preparation; (2) the technique must allow the analysis of a very small amount of sample and, (3) it must be a non-destructive technique. X-ray diffraction [43] and Raman spectroscopic [44] techniques satisfy these criteria. Raman spectroscopy can differentiate among closely related compounds and provides structural information, but the extraction of this information requires a reliable assignment of the main bands in the Raman spectrum and models to interpret the values of the Raman shifts. Raman spectroscopy has been already used to characterize the secondary phases of SNF, but the nuclear Raman database is still under development. The excellent works performed by Frost et al. [45–51] should be underlined.

Theoretical solid-state calculations allow a safe and complete characterization of uranyl-containing materials free of the inconveniences associated to their radiotoxicity [42, 52–63]. However, the application of these methods to the study of mineral phases containing Rare Earth Elements is difficult not only due to the complexity of these materials but also to the high level of theory needed to describe these materials [64, 65]. The number of valence electrons in the outer shells, which should be described explicitly, is very large and occupy high angular momentum orbitals. The inner electrons of these elements must be described by using relativistic pseudopotentials. Since an accurate pseudopotential specific for the uranium atom suitable for the realization of vibrational studies of solids was not available, a new norm-conserving relativistic pseudopotential was developed recently [42, 52]. The use of this pseudopotential has permitted to perform a large series of studies about the crystal chemistry and bonding of the uranium atom in these systems.

The theoretical studies have allowed to confirm the crystal structures of many of these materials determined from X-ray diffraction data by structure refinement, which have never been studied using theoretical methods [54, 55]. The knowledge of the crystal structures of uranyl-containing minerals has experienced a large improvement in the last times [24] due to the use of charge-coupled device (CCD) detectors for X-ray diffraction [66]. These detectors have allowed to perform reliable structural determinations of small size crystals and of materials comprising large unit cells, both cases being frequent for this kind of mineral phases. However, the determination of the hydrogen atom positions in many of these structures has not been possible. Two important examples are schoepite [56] and becquerelite phases [63]. The hydrogen atom positions in these structures were successfully determined theoretically [56, 63]. The calculations were performed using theoretical solid-state methods based on density functional theory using plane waves and pseudopotentials [67].

Similarly, as was noted by Weck et al. [68, 69], the existence of a great amount of information on the formation, thermodynamic stability, and phase transformations of alteration phases formed at the SNF surface is in contrast with the paucity of data regarding the mechanical stability and properties of these phases. This is surprising since the underlying atomistic deformation modes and interactions determine thermodynamic phase stability and transformation. Except the theoretical studies of Weck et al. [68, 69] on the uranyl peroxide hydrates, studtite and metastudtite, and the studies of the mechanical behavior of uranium dioxide [70–72], no experimental or computational studies have reported the mechanical properties of these phases. Furthermore, the Born conditions of mechanical stability of the corresponding structures have not been analyzed. Whereas the lack of the experimental data could be related to the special care needed to handle these radioactive minerals, computational data have not been reported because of the difficulties in the application of theoretical methods to uranium-containing solids. However, from the computed structures, the mechanical properties and stability of rutherfordine, soddyite, uranophane, schoepite and becquerelite mineral phases were reported recently [54–56, 63], based on the computed elasticity tensor. The elastic constant tensor of an inorganic compound provides a complete description of the response of the material to external stresses in the elastic limit [73] and is usually correlated with many mechanical properties such as the bulk and shear moduli, stiffness coefficients and anisotropy factors. The corresponding equations of state were also determined [54–56, 63].

Density Functional Perturbation Theory [74–76] allows the accurate determination of the vibrational Raman spectra with relatively small cost/performance ratios. Thus, although the computations required are generally quite expensive, these methods permit to perform a complete and rigorous assignment of the Raman vibrational bands. Whereas in the experimental works, the assignment is usually performed in an incomplete manner and by using empirical arguments, the theoretical methods provide graphical representations of the vibrational motions of the atoms in the corresponding normal modes. However, due to the lack of a good pseudopotential for uranium atom, there were very few published works on the theoretical vibrational spectra of uranium-containing solids [77–81] and the unique spectral features considered in these studies were the vibrational band wavenumbers.

This chapter provides mainly a review of the calculated structures, X-ray and mechanical properties of uranyl-containing minerals. The computed structures, shown together, display the most common coordination structures of the uranium atom in these minerals and give a clear introduction to the crystal chemistry of uranyl ion. Furthermore, because there were not experimental values for a large fraction of the mechanical properties obtained in our previous works, the theoretical techniques were shown to have a highly predictive value. The data provided by these mechanical property calculations are within the most important mineral information, which is usually given to characterize a given mineral and the values reported together permit to extract the range of values which is likely to find for other uranyl-containing minerals. Finally, an example of the computed vibrational Raman spectrum for soddyite uranyl silicate mineral is provided to show the accuracy of the calculated spectra and how it was used to complement the experimental work in the assignment of the bands of the observed spectrum [54].

## 2. Methods

The generalized gradient approximation (GGA) together with PBE functional [82] supplemented with Grimme empirical dispersion correction [83], were used to study the soddyite, uranophane-α and schoepite mineral phases. The introduction of dispersion corrections improved significantly the computed structural and vibrational properties as a result of the better description of the hydrogen bonding present in the corresponding structures. For the case of rutherfordine mineral, the specialized version of PBE functional for solid materials, PBEsol [84], was used instead. These functionals are implemented in CASTEP program [85], a module of the Materials Studio package [86], which was employed to model the structures of the materials considered. The pseudopotentials used for H, C, O, Si, and Ca atoms in the unit cells of these minerals were standard norm-conserving pseudopotentials [87] given in CASTEP code (00PBE-OP type). The norm-conserving relativistic pseudopotential for U atom was generated from first principles as shown in previous works [42, 52]. Whereas our uranium atom pseudopotential includes scalar relativistic effects, the corresponding pseudopotentials used for H, C, O, Si, and Ca atoms do not include them. This pseudopotential has been used extensively in the research of uranyl-containing materials [42, 52–63].

### 2.1. Crystal structures and mechanical properties

The atomic positions and cell parameters were optimized by using the Broyden-Fletcher-Goldfarb-Shanno method [67, 88] with a convergence threshold on atomic forces of 0.01 eV/Å. The kinetic energy cut-off and k-point mesh [89] were chosen to ensure good convergence for computed structures and energies. The structures of the minerals considered in this work were optimized in calculations with augmented complexity by increasing these parameters. The precise calculation parameters used to determine the final results may be found in the corresponding articles [42, 53–56].

The elastic constants required to calculate the mechanical properties and to study the mechanical stability of the crystal structures were obtained from stress-strain relationships using the finite deformation method. Finite programmed symmetry-adapted strains [73] are used in this method to extract the individual constants from the stress tensor obtained as response of the system to the applied strains. This stress-based method was shown to be more efficient than the energy-based methods and the use of DFPT technique for the calculation of the elasticity tensor [90].

The elastic modulus and the corresponding derivatives with respect to pressure for the mineral phases rutherfordine, soddyite, uranophane-α and schoepite were calculated by fitting lattice volumes and associated pressures to a fourth-order Birch-Murnaghan equation of state (EOS) [91]. The lattice volumes near the equilibrium geometry were determined by optimizing the structure at several applied pressures with values in the range −1.0 and 12 GPa. EOSFIT 5.2 code [92] was used to adjust the results to the selected equation of state.

### 2.2. Vibrational Raman spectra

The vibrational spectra were calculated using the linear response density functional perturbation theory (DFPT) [74–76] implemented in the CASTEP code in the same way as in previous works [42, 54–57, 93, 94]. The phonon frequencies at the gamma point of the Brillouin zone were determined using atomic displacement perturbations. Raman intensities are third-order derivatives of total energy with respect to atomic positions and laser electric field (twice). These are calculated in CASTEP [95] by using a combination of perturbation theory to evaluate the second derivatives with respect to applied fields and finite differences to evaluate the third derivatives with respect to atomic displacement. The frequencies presented in this work have not been scaled to correct for anharmonicity and remaining errors of the theoretical treatment employed, such as incomplete treatment of electron correlation and basis set truncation [96].

# 3. Results and discussion

The unit cell parameters and internal atomic positions were first optimized using initial atomistic models based in the atomic coordinates given by several authors [97–100]. From the optimized structures, we have obtained both the structural parameters as well as the X-ray powder patterns. Then, the mechanical properties, the EOS, the vibrational Raman spectrum and thermodynamic and optic properties were determined [42, 53–56].

## 3.1. Crystal structures

### *3.1.1. Rutherfordine*

Christ et al. [101, 102] presented two possible orthorhombic structures for rutherfordine. The first is consistent with Pmmn symmetry and the second with Imm2 symmetry. The structure was later refined by Finch et al. [97]. We considered both orthorhombic structures for rutherfordine. The results found were nearly the same and the energy difference for the optimized structures was less than 0.001 eV. Therefore, only the results obtained for the Imm2 structure are described in detail here. The results for the Pmmn structure may be found in Ref. [42]. The results encountered [42] pointed that both structures may be simultaneously present in nature in accordance with the suggestion of Christ et al. [101, 102].

The computed structure of rutherfordine is shown in **Figure 1A** and **B**. It contains approximately linear $UO_2^{2+}$ uranyl ions that are coordinated by six O atoms arranged at the equatorial vertices of uranyl hexagonal bipyramids. These O atoms belong to four carbonate ligands, and U is bonded with two carbonate ions in a bidentate manner and two in a monodentate manner. Each uranyl polyhedron is linked to two other uranyl polyhedra in a trans arrangement by edge sharing, resulting in chains of polyhedra. Adjacent chains are linked by the sharing of equatorial vertices between uranyl polyhedra, which results in a sheet structure that contains triangular voids. Carbonate triangles occupy one half of the voids, such that they share the equatorial edges of two adjacent uranyl hexagonal bipyramids and single vertices of two additional uranyl polyhedra (see **Figure 1A**). The resulting sheets or layers are electroneutral, and adjacent sheets in rutherfordine are bonded together by van der Waals forces. The $UO_2CO_3$ layers are staggered with respect to the layer above or below, such that uranyl units lie above and below a carbonate carbon atom in adjacent layers. Layers are separated by a distance of about 4.6 Å. The structure is similar for Imm2 and Pmmn structures. For the Pmmn structure the carbonate triangles in contiguous sheets point in opposite directions while in the Imm2 one they point in the same direction (see **Figure 1B**, where two contiguous sheets in the structure of rutherfordine are shown).

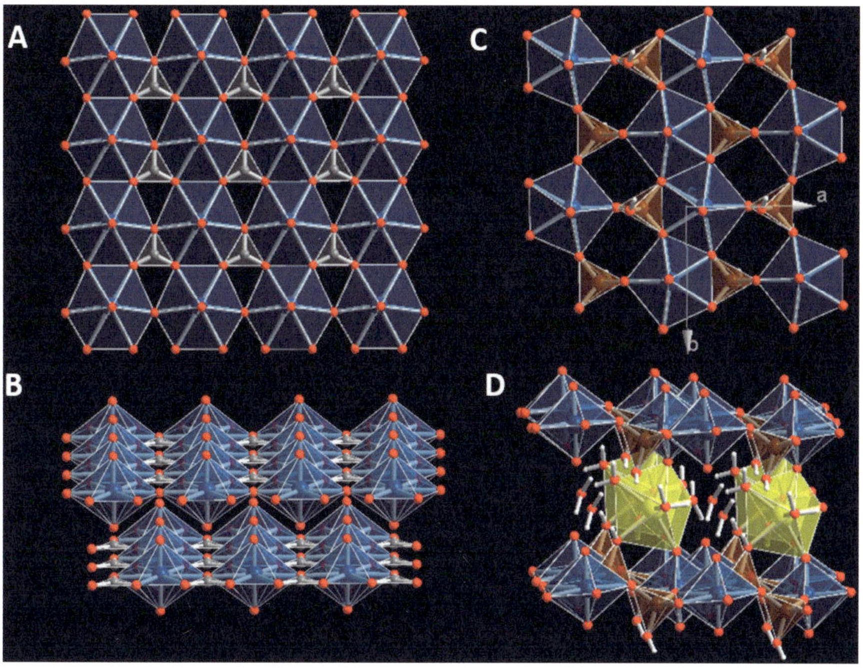


**Figure 1.** Structures of rutherfordine and uranophane-α minerals; (A) view of a sheet in the structure of rutherfordine from [010]; (B) two contiguous sheets in the structure of rutherfordine; (C) view of a 2 × 2 × 2 supercell of uranophane-α from [001] showing a uranyl silicate sheet composed of $UO_7$ pentagonal bipyramids and $SiO_3OH$ tetrahedra; (D) view of a 2 × 2 × 2 supercell from [010] showing two uranyl silicate sheets connected by $CaO_2(OH)(H_2O)_4$ polyhedra. One water molecule is out of the Ca coordination polyhedra (free water). Color code: U—blue, O—red, C—gray, Ca—yellow, H—white.

### *3.1.2. Uranophane-α*

The unit cell of uranophane-α has monoclinic symmetry [98]. Stohl and Smith [103] categorized naturally occurring uranyl silicates according to the uranium to silicon ratio, which in part determines the structures of these minerals. Most uranyl silicate minerals have 1:1 uranium to silicon ratio and are sheet silicates [27, 28, 98, 103]. Uranyl silicate sheets are composed of $[(UO_2)_2(SiO_4)_2]^{-4}$ units bound at the equatorial edges (see **Figure 1C**). A sheet, $[(UO_2)(SiO_4)]_n^{-2n}$, contains $UO_7$ pentagonal bipyramids and $SiO_3OH$ tetrahedra. Charge compensating cations, calcium in uranophane, lie in the interlayer space between the sheets (see **Figure 1D**). Two uranyl silicate sheets are connected by $CaO_2(OH)(H_2O)_4$ polyhedra (distorted pentagonal bipyramid). The Ca atom ligands are four water molecules, two uranyl oxygen atoms belonging to the upper and the lower sheets, and one OH group of $SiO_3OH$ tetrahedra. One water molecule is out of the Ca polyhedra. While this water molecule is described as free or crystallization water, the four water molecules belonging to the Ca atom coordination sphere are referred to as structural water. Hydrogen bonds reinforce the bonding between the uranyl silicate sheets, the Ca atom, and the free water. As it can be seen in **Figure 1D**, the upper sheet $SiO_3OH$ tetrahedra have free OH groups pointing downwards and the lower sheet tetrahedra have OH groups pointing upwards, which belong to the Ca atom coordination sphere.

### 3.1.3. Soddyite

While most uranyl silicate minerals have 1:1 uranium to silicon ratio and are sheet silicates [27, 28, 103], soddyite exhibits a 2:1 uranium to silicon ratio and framework crystal structure [99]. The computed structure is shown in **Figure 2A** and **B**. **Figure 2A** shows a view of the unit cell from [001]. **Figure 2B** is a view of a 2 × 2 × 1 supercell along [110] where only a subset of atoms is retained in order to show a clearer view of the soddyite structure. As in the case of the other uranyl silicate considered, uranophane-$\alpha$, U atoms display pentagonal bipyramid coordination, $UO_6(H_2O)$, and Si atoms present tetrahedral coordination, $SiO_4$. The U bipyramids are connected by sharing two non-adjacent edges of the equatorial plane to form zigzag chains (see **Figure 2B**). The single unshared equatorial vertex of the bipyramid is occupied by $H_2O$. All the water molecules are crystallization water, that is, belong to the coordination structure of uranium atom. The chains are parallel to [110] plane and are cross bonded through two opposite edges of the $SiO_4$ tetrahedra; i.e. adjacent uranyl silicate chains are directly linked as each tetrahedron shares two edges with bipyramids from two different chains. Moreover, the cohesion of the structure is enhanced by a pattern of hydrogen bonds involving the water molecules and the uranyl O atoms.

### 3.1.4. Schoepite

The computed structure of schoepite is displayed in **Figure 2C** and **D**. There are eight symmetrically distinct uranium sites in the structure of this mineral [100] (see **Figure 2D**). All U atoms are coordinated by seven anions in pentagonal bipyramidal arrangements. Each

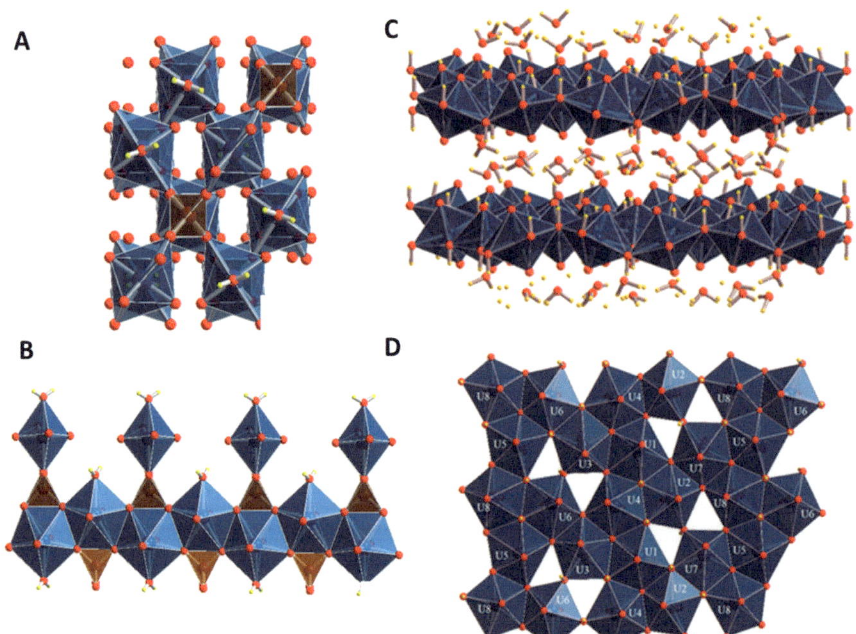


**Figure 2.** Structure of soddyite and schoepite minerals; (A) view of the unit cell of soddyite from [001]; (B) view of a 2 × 2 × 1 supercell of soddyite from [110], retaining only a subset of atoms (a single chain of U bipyramids cross bonded to other chains through two opposite edges of the $SiO_4$ tetrahedra is shown); (C) view of the full unit cell of schoepite from [100]; and (D) view of a schoepite sheet from [001] direction. Color code: U—blue, Si—brown, O—red, H—yellow.

pentagonal bipyramid, referred to as $UO_7$ below, consists of two apical $O^{2-}$ anions and five equatorial anions, $O^{2-}$ or $OH^-$. While the uranium atoms U2 and U6 (see **Figure 2D**) have a coordination environment of $UO_2(OH)_5$, the coordination for the remaining uranium atoms is $UO_2O(OH)_4$. The most stable configuration around a uranyl, $UO_2^{2+}$ group has a pentagonal arrangement of equatorial anions as predicted for uranyl oxyhydroxides by Evans in 1963 [104]. The $UO_7$ pentagonal bipyramids share edges to form dimers, which in turn, link by sharing edges to form staggered ribbons along the [100] direction. Then, these ribbons cross-link in the [010] direction by sharing edges and corners of the polyhedra. This results in a strongly bonded sheet of stoichiometry $[(UO)_8O_2(OH)_{12}]$ parallel to plane {001}. This sheet constitutes the structural unit of schoepite, which stacks along the [001] direction. As the sheets are neutral, they are linked together by H-bonding only, through a complex network of H-bonding involving interlayer $H_2O$ groups and $O^{2-}$ or $OH^-$ groups in the structural sheet.

The structure within the uranium oxide hydroxide layers is basically the same in schoepite [100] metaschoepite [105] and many other uranium(VI) complex oxides, such as fourmarierite [106]. While most equatorial $O^{2-}$ and $OH^-$ groups are shared as vertices of three $UO_7$ polyhedra, the O atoms in some hydroxyl groups bridge two uranyl polyhedra alone [105]. The corresponding sections of the layers not occupied by the pentagons of the $UO_7$ groups have a bow-tie-like motif (see **Figure 2D**) connected centrally at the bridging hydroxyl. There are 12 $H_2O$ groups in schoepite interlayer. Ten of these $H_2O$ groups are located at the apices of two distorted pentagons and the remaining two water molecules are not members of the pentagonal rings and are located between the pentagonal rings. The pentagonal rings vertices are nearly at the positions of the equatorial anions in the U1 and U7 polyhedra from the two adjacent sheets. The general hydrogen bond structure described by Finch et al. [100] was properly reproduced by present theoretical calculations, confirming the suggested hydrogen bond structure. Besides, our results provided the locations for the hydrogen atoms in the full unit cell, which were never precisely obtained by either theoretical or experimental methods.

#### *3.1.5. Unit cell parameters*

As it has been mentioned in Section 2, the structures of these materials were determined in calculations with augmented complexity. **Table 1** gives the final lattice parameters, volumes and densities obtained compared with the experimental ones [97–100]. As it can be seen, the theoretical calculations reproduce the experimental information accurately. The errors in the computed volumes and densities with respect to experimental data were very small, 0.1, 0.3, 0.4 and 2.1% for rutherfordine, soddyite, uranophane-$\alpha$ and schoepite, respectively.

### 3.2. X-ray powder patterns

The X-ray powder diffractograms of these minerals were calculated from the computed structures using $CuK_\alpha$ radiation ($\lambda$ = 1.540598 Å). It must be outlined that the computation of the X-ray powder pattern of a given material does not require any additional optimization or response calculation. Their determination may be made directly from the optimized atomic positions and unit cell parameters [107]. The calculated patterns are compared in **Figure 3** with the X-ray diffractograms computed from the experimental geometries of rutherfordine, soddyite and uranophane-$\alpha$ [97–99] (**Figure 3A–C**). The comparison of the

| Material | Source | $a$ (Å) | $b$ (Å) | $c$ (Å) | α | β | γ | Vol. (Å$^3$) | Dens. (g/cm$^3$) |
|---|---|---|---|---|---|---|---|---|---|
| Rutherfordine | Calc. [42] | 4.8267 | 9.3639 | 4.2727 | 90 | 90 | 90 | 193.1 | 5.675 |
| | Exp. [97] | 4.840 | 9.273 | 4.298 | 90 | 90 | 90 | 192.9 | 5.682 |
| Soddyite | Calc. [54] | 8.0780 | 11.4253 | 18.8380 | 90 | 90 | 90 | 1738.6 | 5.104 |
| | Exp. [99] | 8.334 | 11.212 | 18.668 | 90 | 90 | 90 | 1744.4 | 5.088 |
| Uranophane-α | Calc. [55] | 6.6689 | 7.0022 | 15.8684 | 90 | 98.07 | 90 | 733.7 | 3.8766 |
| | Exp. [98] | 6.665 | 7.002 | 15.909 | 90 | 97.27 | 90 | 736.5 | 3.8618 |
| Schoepite | Calc. [56] | 14.2740 | 16.8076 | 14.4841 | 90 | 90 | 90 | 3474.9 | 4.993 |
| | Exp. [100] | 14.337 | 16.813 | 14.731 | 90 | 90 | 90 | 3550.9 | 4.886 |

**Table 1.** Lattice parameters of rutherfordine, soddyite, uranophane-α and schoepite.

patterns derived directly from the structures is free of all possible interferences since both are determined under identical conditions. The agreement in line positions and intensities is very good. Nevertheless, the direct comparison of the calculated and experimental patterns was also very good [42, 54, 55]. Program XPowder [108] using the PDF-2 database [109] recognized the calculated spectra as those of the corresponding materials using low tolerance limits for the spectral differences. In the case of schoepite mineral, the X-ray powder spectrum was computed using REFLEX code included in Materials Studio package [86] and the results are compared in **Figure 3D** with the experimental X-ray diffractogram from the record 100188 of the RRUFF database [110] which corresponds to a natural schoepite mineral from Shinkolobwe mine, Katanga, Congo. The agreement of the computed and experimental diffractograms is excellent.

### 3.3. Mechanical properties

#### *3.3.1. Mechanical stability*

The elastic tensor, needed for the calculation of mechanical properties and to study the mechanical stability of a crystal structure, was calculated at the optimized equilibrium structure from stress-strain relationships, by using the finite deformation method implemented in CASTEP program. Materials with orthorhombic (rutherfordine, soddyite and schoepite) and monoclinic (uranophane-α) unit cells have 9 and 13 non-degenerate elastic constants, C(i,j) in the symmetric stiffness matrix [68, 73], respectively. The computed values of these constants are given in Refs. [53–56].

For orthorhombic systems, a set of necessary and sufficient Born [111, 112] conditions for mechanical stability are known [68, 113]. These conditions can be expressed as a set of algebraic inequalities among products of elastic constants and were properly satisfied by the computed stiffness tensors of rutherfordine, soddyite and schoepite [53, 54, 56]. For monoclinic crystals, a set of necessary (but not sufficient) Born criteria for mechanical stability can also be written in terms of the stiffness matrix elements [68]. These conditions were fully satisfied by the computed elastic tensor of uranophane-α [55]. The generic necessary and sufficient

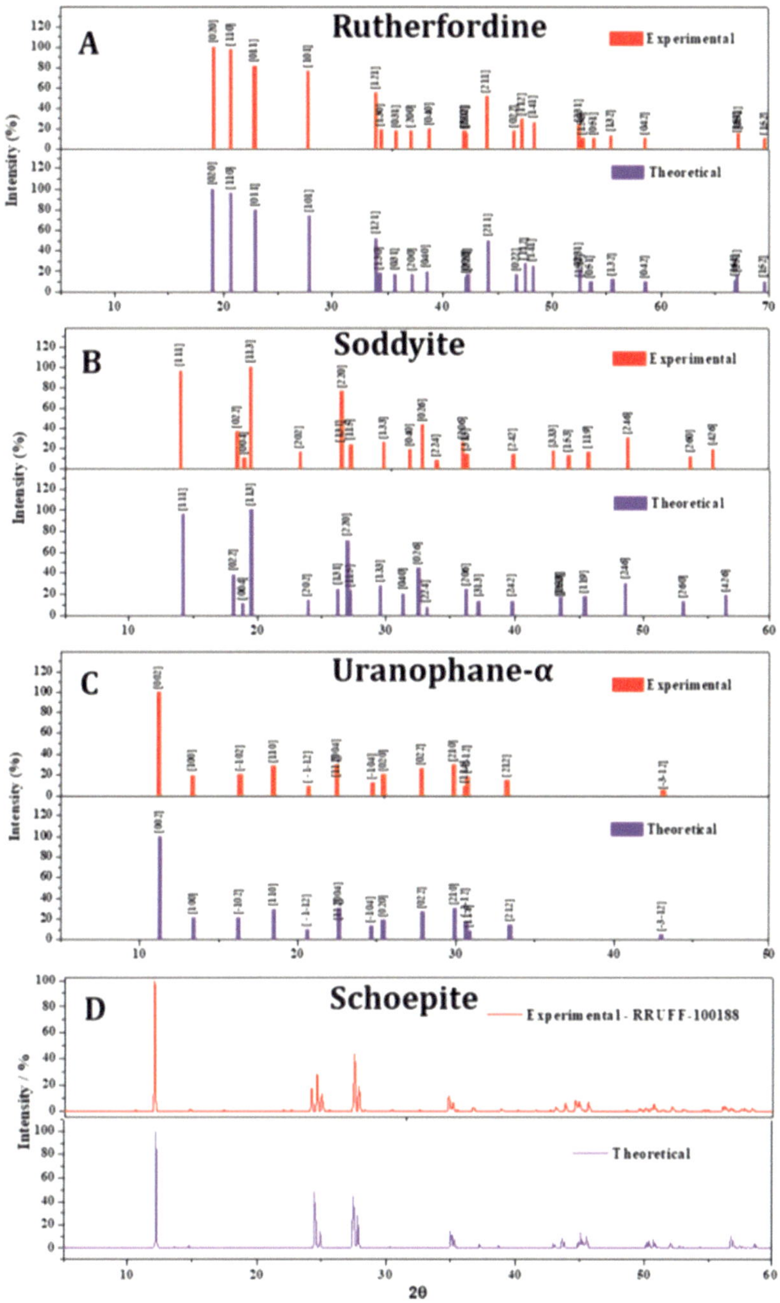


**Figure 3.** X-ray powder patterns of the materials under study using $CuK_{\alpha}$ radiation: (A) rutherfordine; (B) soddyite; (C) uranophane-$\alpha$; and (D) schoepite.

Born criterion of stability is that all eigenvalues of the C matrix be positive [113]. The C matrix was diagonalized numerically, and all eigenvalues were found to be positive. Since the Born conditions were satisfied by the computed elastic tensors of all the materials studied, their mechanical stability was inferred [53–56].

The dynamical stability should also be analyzed to study the stability of the material in a complete form. A necessary and sufficient condition for the dynamical stability of a structure is that all of its phonon modes have positive frequencies for all wave vectors [113]. Since the fulfillment of this condition was also verified from the phonon calculations employed to determine the thermodynamic properties of these materials [53, 59, 62], their crystal structures were found to be mechanically and dynamically stable.

If the stiffness matrix of a material has a diagonal element associated to a certain direction which is much smaller than the other ones, it may be suggested that the thermal expansion of the material will occur predominantly along this direction. This feature was found to occur in the case of rutherfordine, uranophane-$\alpha$ and schoepite [53, 55, 56] all of them being layered uranyl-containing minerals, along the direction perpendicular to the sheets characterizing their structures. This is consistent with the fact that the intersheet space in layered materials increase to a large extent as temperature increases due to the fact that the sheets are generally bonded only by weak van der Waals forces (rutherfordine [53]) or by hydrogen bonding among the sheets and water molecules present in their interlayer space (uranophane-$\alpha$ and schoepite [55, 56]).

### *3.3.2. Mechanical properties*

If single crystal samples are not available, the measurement of the individual elastic constants is not possible. Instead, the polycrystalline bulk modulus (B) and shear modulus (G) may be determined experimentally. The Voigt [114] and Reuss [115] schemes were used to compute the isotropic elastic properties of polycrystalline aggregates of these materials [53–56]. In Voigt method for calculating the elastic moduli, the strain throughout the aggregate of crystals is considered uniform, and the relations expressing the stress are averaged over all possible lattice orientations. While the strain is assumed to be uniform throughout the aggregate of crystals in Voigt's method, Reuss approximation considers the stress to be uniform and the averaging of the relations expressing the strain is carried out. As shown by Hill [116], the Reuss and Voigt approximations result in lower and upper limits, respectively, of polycrystalline constants and practical estimates of the polycrystalline bulk and shear moduli in the Hill approximation can be computed using average formulas. The formulae for these approximations may be found in several sources [68, 117]. Although the differences between the results obtained for rutherfordine, soddyite and schoepite in the Reuss and Voigt approximations was generally small, this difference was found to be quite large for rutherfordine. This reason for this behavior is that rutherfordine is a highly anisotropic material showing large differences between the values of the elastic constants along different directions [53, 68].

The Reuss scheme provided the best results when the computed bulk moduli were compared with that determined from the equation of state (EOS) as it occurred in other works by other authors [118, 119]. In the case of uranophane-$\alpha$, the Hill approximation gave the best results [55]. The values of the mechanical properties computed in in the Reuss approximation for rutherfordine, soddyite and schoepite and in the Hill approximation for uranophane-$\alpha$, are given in **Table 2**. CASTEP code gave a numerical estimate of the error in the computed bulk moduli, B, of 0.94, 2.31, 2.45 and 2.28 GPa for rutherfordine, soddyite, uranophane and schoepite, respectively, and, consequently, our final values of the bulk moduli computed from the elastic constants are, 17.97 ± 0.94, 58.41 ± 2.31, 59.20 ± 2.45 and 34.53 ± 2.28 GPa, respectively.

| Property | Rutherfordine | Soddyite | Uranophane-α | Schoepite |
|---|---|---|---|---|
| B | 17.97 | 58.41 | 59.20 | 34.53 |
| G | 19.47 | 36.00 | 36.52 | 23.17 |
| E | 42.92 | 89.60 | 90.88 | 56.80 |
| ν | 0.10 | 0.24 | 0.24 | 0.23 |
| D | 0.92 | 1.62 | 1.62 | 1.49 |
| H | 9.47 | 6.24 | 6.33 | 4.88 |

**Table 2.** Computed bulk, modulus, shear modulus, Young modulus, Poisson ratio, Pugh's ratio, and Vickers hardness (B, G, E, ν, D, and H) for rutherfordine, soddyite, uranophane-α and schoepite [53–56].

While the elasticity theory is very well understood and mathematically well founded, it is difficult to visualize how the elastic properties vary with the strain orientation, except for the simplest cases of isotropic materials. In order to address this difficulty, the ElAM software of Marmier et al. [120] was used to obtain detailed tridimensional representations of the most important elastic properties, which are shown in **Figure 4** for schoepite [56]. In **Figure 4A**, the property displayed is the inverse of the bulk modulus (the compressibility) instead of the bulk modulus. As it can be seen in **Figure 4A**, the vertical direction (c axis) is the most compressible one in accordance with the previous discussion on the results of the stiffness matrix C matrix. Also, it must be noted that the corresponding tridimensional representations of the elastic properties of metaschoepite mineral, including those of the shear modulus, are very similar to those reported in **Figure 4**. This means that although the dehydration from schoepite to metaschoepite leads to a change of space symmetry [97, 105] the transformation is not shear induced, as occurs for the dehydration of studtite to metastudtite [69]. This behavior is the expected one, since the structures of schoepite and metaschoepite are very similar, the main changes being the differences in the arrangements of the interlayer water molecules and associated hydrogen bonds [105].

In general, a large value of shear moduli is an indication of the more pronounced directional bonding between atoms. Shear modulus represents the resistance to plastic deformation while the bulk modulus represents the resistance to fracture [117, 121]. Young modulus defines the relationship between stress (force per unit area) and strain (proportional deformation) in a material, that is, $E = \sigma/\varepsilon$. Pugh [122] introduced the proportion of bulk to shear modulus of polycrystalline phases ($D = B/G$) as a measure of ductility from the interpretation of the shear and bulk modulus given above. The value separating ductile and brittle materials is 1.75, i.e., if $D > 1.75$, the material behaves in a ductile manner, otherwise the material behaves in a brittle manner [68]. The Poisson ratio, ν, can be also utilized to measure the malleability of crystalline compounds, and is closely related to the Pugh's ratio. The Poisson ratio is close to 0.33 (1/3) for ductile materials, while it is generally much less than 0.33 for brittle materials. As it can be seen for rutherfordine, soddyite, uranophane and schoepite we find ratios D of 0.92, 1.62, 1.62, and 1.49, respectively. Similarly, the calculated Poisson ratios, ν are 0.10, 0.24, 0.24, 0.23, respectively. These values are smaller than 1.75 (D) and 0.33 (ν), corresponding to brittle materials. For comparison, studtite and metastudtite uranyl peroxide minerals were found to be ductile [68].

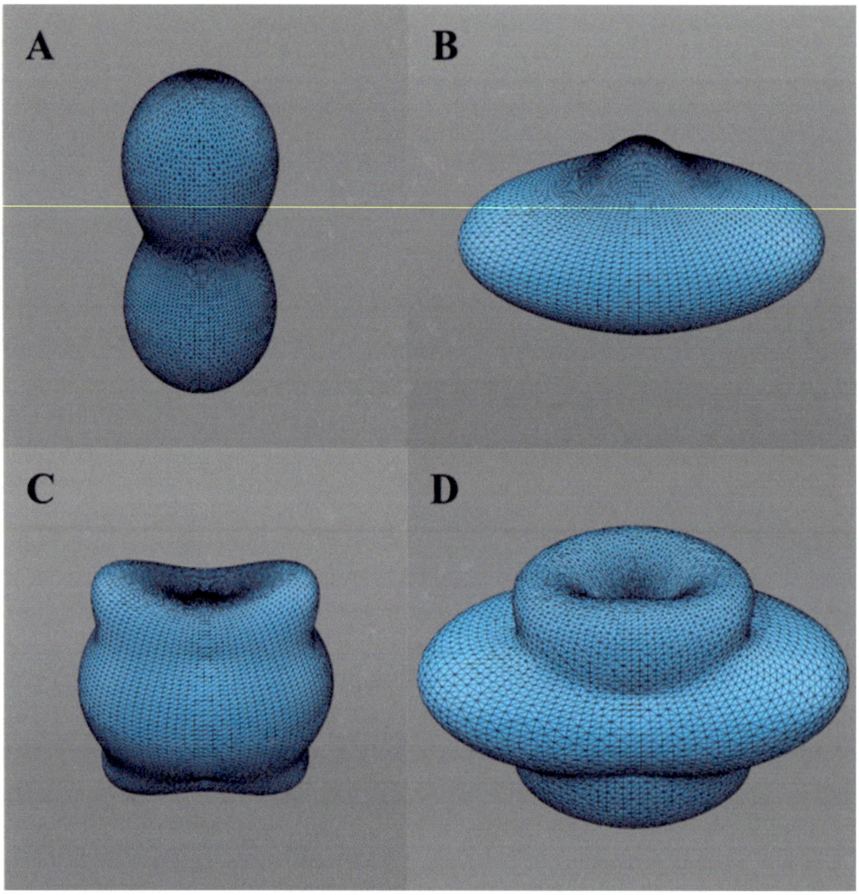


**Figure 4.** Schoepite elastic properties as a function of the orientation of the applied strain: (a) compressibility; (b) young modulus; (c) shear modulus; and (d) Poisson ratio.

Hardness of these systems is computed according to a recently introduced empirical scheme [123], which correlates the Vickers hardness and the Pugh's ratio (D = B/G). Vickers hardness, H, values of polycrystalline rutherfordine, soddyite, uranophane and schoepite are given in **Table 2**. Rutherfordine is a material characterized by a quite large hardness, 9.5, and the three other materials have intermediate hardness values 6.2, 6.2, and 4.8. For comparison, we obtained the hardness of studtite and metastudtite [52] using the elasticity data of Weck et al. [68]. These systems, characterized by much larger D ratios, have much smaller hardness (smaller than one). The two uranyl silicates studied, uranophane and soddyite, have very similar hardness values of about 6.2 [54, 55].

In order to assess the elastic anisotropy of these minerals, shear anisotropic factors were obtained. These factors provide a measure of the degree of anisotropy in the bonding between atoms in different planes and are very important to study the material durability [117, 121]. Shear anisotropic factors for the {100} ($A_1$), {010} ($A_2$), and {001} ($A_3$) crystallographic planes were computed. For an isotropic crystal, the factors $A_1$, $A_2$, and $A_3$ must be one, while any value smaller or greater than unity is a measure of the degree of elastic anisotropy possessed by the crystal. For example, for uranophane [55], the computed anisotropy factors were $A_1 = 0.44 < A_2 = 0.51 < A_3 = 0.64$ and, consequently, the {100} plane is shown to be the most anisotropic one.

The recently introduced universal anisotropy index [124], $A^U$, is another important measure of crystal anisotropy. The departure of $A^U$ from zero defines the extent of single crystal anisotropy and accounts for both the shear and the bulk contributions unlike all other existing anisotropy measures. Thus, $A^U$ represents a universal measure to quantify the single crystal elastic anisotropy. Rutherfordine, soddyite, uranophane, schoepite studtite and metastudtite are characterized by universal anisotropy indices of 8.81, 0.50, 0.81, 0.78, 2.17 and 1.44 [53–56, 68]. Therefore, while rutherfordine is strongly anisotropic [53] and studtite and metastudtite have quite large anisotropies [68], soddyite, uranophane and schoepite have very small anisotropies [54–56] ($A^U = 0$ corresponds to a perfectly isotropic crystal).

A set of fundamental physical properties may be estimated with the calculated elastic constants. For example, $V_L$ and $V_T$, the transverse and longitudinal elastic wave velocities in the polycrystalline materials may be determined in terms of the bulk and shear moduli [68]. The values obtained are presented in **Table 3**.

### 3.3.3. *Equations of state*

Unit cell volumes were determined by calculating the optimal structures at 17 different applied pressures between −1.0 and 12.0 GPa. The computed volume and pressure values were fitted to fourth-order Birch-Murnaghan [91] equations of state (EOS) by employing the EOSFIT 5.2 program [92]. The values found for bulk modulus and its first and second derivatives, respectively, at the temperature of 0 K (B, B′, and B″) are given in **Table 4**. The corresponding values of the

| Velocity component | Rutherfordine | Soddyite | Uranophane | Schoepite | Studtite | Metastudtite |
|---|---|---|---|---|---|---|
| VT (km/s) | 2.367 | 2.708 | 3.069 | 2.217 | 1.74 | 1.69 |
| VL (km/s) | 3.820 | 4.671 | 5.276 | 3.759 | 3.31 | 3.50 |

**Table 3.** Transverse and longitudinal elastic wave velocities of rutherfordine, soddyite, uranophane, schoepite studtite and metastudtite [53–56, 68].

| Property | Rutherfordine | Soddyite | Uranophane | Schoepite |
|---|---|---|---|---|
| B (GPa) | 19.03 ± 0.36 | 60.07 ± 0.67 | 59.96 ± 2.1 | 35.17 ± 0.39 |
| B′ | 15.34 ± 0.72 | 4.19 ± 0.60 | 2.29 ± 1.11 | 7.39 ± 0.40 |
| B″ ($GPa^{-1}$) | −7.43 ± 1.32 | 0.25 ± 0.20 | −0.25 ± 0.19 | −1.31 ± 0.22 |

**Table 4.** Bulk modulus and its first and second derivatives at the temperature of 0 K (B, B′, and B″) for rutherfordine, soddyite, uranophane and schoepite [53–56] determined from fits to a fourth-order Birch-Murnaghan EOS.

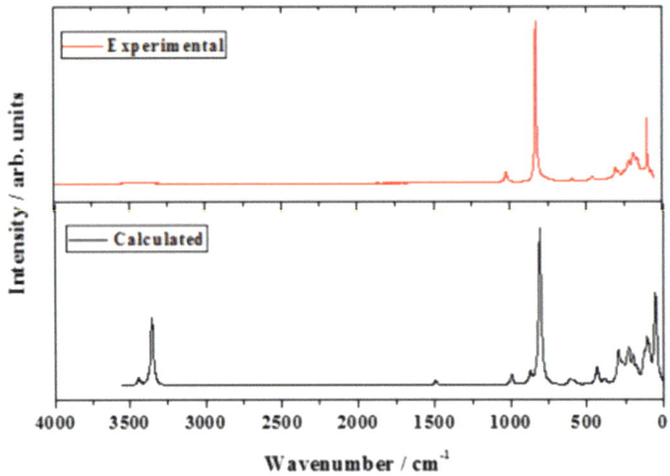


**Figure 5.** Experimental and calculated Raman spectra of soddyite mineral [54].

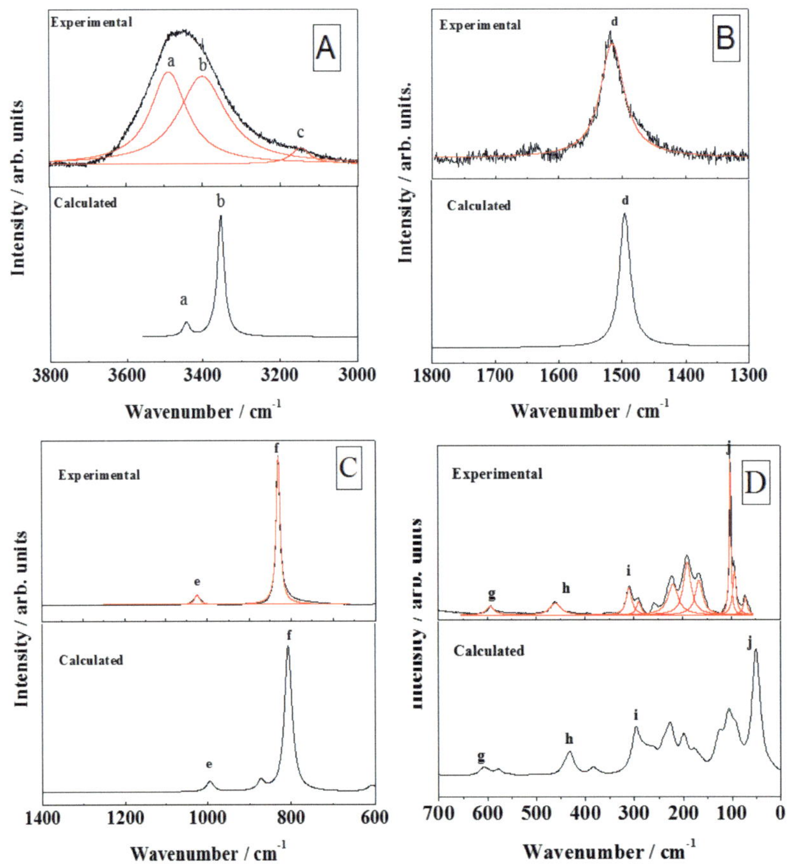


**Figure 6.** Experimental and theoretical Raman spectra of soddyite mineral. (A) Region: 3800–3000 $cm^{-1}$; (B) region: 1800–1300 $cm^{-1}$; (C) region: 1400–700 $cm^{-1}$; (D) region: 700–0 $cm^{-1}$.

bulk modulus of rutherfordine, soddyite, uranophane and schoepite obtained from the elastic constants are 17.97 ± 0.94, 58.41 ± 2.31, 59.20 ± 2.45 and 34.53 ± 2.28 GPa, respectively [53–56].

### 3.4. Vibrational Raman spectra

The vibrational Raman spectrum of soddyite was computed using DFPT and fully assigned [54]. In this section, we will provide a brief resume of the results obtained in order to show a representative example of the theoretical procedure used to characterize this mineral phase and assign the main bands in its Raman spectrum.

The Raman spectrum recorded in the wavenumber range of 3600–0 $cm^{-1}$ is compared with the one obtained theoretically in **Figure 5** and, as it may be observed, the calculated spectrum resembles closely the experimental one. The calculated spectrum was determined at ambient temperature using a laser radiation wavelength of 532 nm and a full width at half maximum of 20 $cm^{-1}$.

The number of contributions of a given band in the experimental spectrum was obtained by using the second derivative method [42, 54]. In **Figure 6**, the experimental and theoretical Raman spectra are displayed in four zones: (A) OH stretching vibration region from 3800 to 3000 $cm^{-1}$ (**Figure 6A**); (B) $H_2O$ bending region 1800–1300 $cm^{-1}$ (**Figure 6B**); (C) uranyl $UO_2^{2+}$ and silicate $SiO_4^{4-}$ fundamental vibrations region from 1400 to 700 $cm^{-1}$ (**Figure 6C**); and (D) low wavenumber region from 700 to 0 $cm^{-1}$ (**Figure 6D**). The wavenumbers of both spectra and the computed intensities and assignments are given in **Table 5**. The Raman shift values and assignments performed by Frost et al. [45, 48] are also provided in the table. The results obtained in each region are described separately in what follows.

**a.** OH stretching vibrations region. In this region, we found one broad band with two contributing bands placed at about 3488 and 3398 $cm^{-1}$. The corresponding calculated Raman shifts were 3443 and 3353 $cm^{-1}$. These two bands were attributed to antisymmetric and symmetric water stretching vibrations, respectively. Although the difference of computed and experimental shifts is quite large, it should be noted that the infrared OH stretching frequencies determined for isolated uranyl silicate clusters have much larger errors in comparison with the experimental data [125]. The low intensity band at wavenumber 3147 $cm^{-1}$ is not found in the calculated spectrum. This band was also found by Frost et al. [46] at 3158 $cm^{-1}$ and it is probably an overtone band ($2\nu_1$, $\nu_1 = 1584$ $cm^{-1}$) [93, 94].

**b.** $H_2O$ bending vibration region. The wavenumber obtained for the water bending vibration, was found at about 1584 $cm^{-1}$, comparable to the computed value of 1495 $cm^{-1}$. Frost et al. [45, 48] encountered an additional shoulder placed at 1596 $cm^{-1}$. This was ascribed to water absorbed on the sample surface [54].

**c.** Uranyl $UO_2^{2+}$ and silicate $SiO_4^{4-}$ fundamental vibration regions. The experimental band at 1024 $cm^{-1}$, is associated to the one calculated at 995 $cm^{-1}$, which was assigned to the $SiO_4^{4-}$ asymmetric stretching vibration. This vibration is shown in **Figure 7A**. Similar values for the wavenumber of this band were found by Frost et al. [45, 48] and Biwer et al. [126] (1025 and 1018 $cm^{-1}$, respectively). The most intense band in the observed Raman spectrum is located about 830 $cm^{-1}$ and computed at 807 $cm^{-1}$. This band was found to be placed at 824 and 828 $cm^{-1}$ by Biwer et al. [126] and Frost et al. [48], respectively. As it can be observed in **Figure 7B**, it must be assigned to uranyl symmetric stretching vibrations. There are two very close bands in the theoretical spectrum at 874 and 873 $cm^{-1}$. Frost et al. [48] also encountered a pair of bands in this region at 909 and 897 $cm^{-1}$, which were assigned to uranyl symmetric stretching vibrations. From in our theoretical results, we believe that the band at 909 $cm^{-1}$ should be associated to the computed band at 873 $cm^{-1}$, which is assigned to symmetric stretching silicate vibrations. Frost et al. [48] also found a band at 791 $cm^{-1}$, which was assigned it to water librational vibrations. This band, however, is close to the computed one at 799 $cm^{-1}$, which is attributed to uranyl symmetric stretching vibrations.

**d.** *Low wavenumber region.* The theoretical bands placed at 610 and 579 $cm^{-1}$ can be compared with the experimental one located at 592 $cm^{-1}$. These bands are attributed to water librational vibrations (twisting and rocking, respectively). Frost et al. [48], assigned the 591 $cm^{-1}$

| Band name | Exp. Raman shift ($cm^{-1}$) [this work] | Exp. Raman shift ($cm^{-1}$) Frost et al. [48] shift/assignment | Calc. Raman shift ($cm^{-1}$) | Irr. rep. (D2h) | Int. (Å⁴) | Assignation |
|---|---|---|---|---|---|---|
| **OH stretching region** | | | | | | |
| *a* | 3488 | 3516/ν(OH) | 3443 | B2g | 3229.1 | $\nu^a$(OH) |
| *b* | 3398 | 3414/ν(OH) | 3353 | Ag | 27818.8 | $\nu^s$(OH) |
| *c* | 3147 | 3158/ν(OH) | — | — | — | — |
| **$H_2O$ bending region** | | | | | | |
| *d* | 1584 | 1584, 1596/δ($H_2O$) | 1495 | Ag | 433.8 | δ($H_2O$) |
| **$UO_2^{2+}$ and $SiO_4^{4-}$ fundamental vibrations region** | | | | | | |
| *e* | 1024 | 1025/$\nu^a$($SiO_4^{4-}$) | 995 | B1g | 750.1 | $\nu^a$($SiO_4^{4-}$) |
| | — | 909, 897/$\nu^a$($UO_2^{2+}$) | 874 | B2g | 53.5 | $\nu^a$($UO_2^{2+}$) + ϱ($H_2O$) |
| | | | 873 | Ag | 583.7 | $\nu^s$($SiO_4^{4-}$) |
| *f* | 830 | 838, 828, 820/$\nu^s$($UO_2^{2+}$) | 807 | Ag | 8054.0 | $\nu^s$($UO_2^{2+}$) |
| | — | 791/ℓ($H_2O$) | 799 | B1g | 387.3 | $\nu^s$($UO_2^{2+}$) |
| **Low wavenumber region** | | | | | | |
| *g* | 592 | 591/δ($SiO_4^{4-}$) | 610 | Ag | 168.6 | t($H_2O$) |
| | | | 579 | B2g | 138.4 | ϱ($H_2O$) |
| *h* | 460 | 459/δ($SiO_4^{4-}$) | 431 | Ag | 323.3 | δ($SiO_4^{4-}$) |
| *i* | 289 | 310/— | 299 | B1g | 54.8 | T($SiO_4^{4-}$) |
| | | | 296 | Ag | 39.6 | t($SiO_4^{4-}$) |
| | | | 295 | B3g | 308.3 | ϱ($SiO_4^{4-}$) |
| *j* | 103 | 111, 102/— | 50 | B2g | 54.5 | $\delta^{op}$(U-$OH_2$) |

**Table 5.** Experimental and calculated Raman band wavenumbers, calculated intensities and assignments. Raman shifts and assignments performed by Frost et al. [48] are also given in the table.

http://dx.doi.org/10.5772/

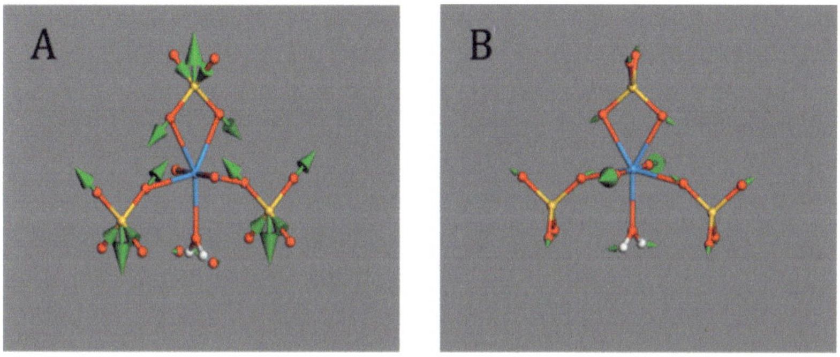


Mode ν=995 cm$^{-1}$ – $\nu^a(SiO_4^{4-})$ – Asymmetric $(SiO_4^{4-})$ stretching.

Mode ν=807 cm$^{-1}$ – $\nu^s(UO_2)^{2+}$– Symmetric $(UO_2)^{2+}$ stretching

**Figure 7.** Pictures of the vibrational motions of the atoms is some Raman active normal modes of soddyite [54].

band to silicate bending vibrations. This last type of vibration is found in the computed spectrum at the wavenumber of 431 cm$^{-1}$, which can be compared to the observed band at 460 cm$^{-1}$. The wavenumbers of this band encountered by Biwer et al. [126] and Frost et al. [48], were 457 and 459 cm$^{-1}$, respectively. While Biwer et al. [126], assigned this band to equatorial uranium-oxygen stretching vibrations, Frost et al. [48], ascribed it to silicate bending vibrations. The last assignment agrees with our assignment. The free silicate ion value for this vibration is 527 cm$^{-1}$ [54]. The theoretical bands situated at 299, 296 and 295 cm$^{-1}$ were mainly assigned to a silicate translation the first, and the other ones to different silicate deformation vibrations (twisting and rocking). They can be compared to the observed band placed at 289 cm$^{-1}$. Finally, the low wavenumber theoretical band situated at 50 cm$^{-1}$ can be approximately mapped to the experimental shift of 103 cm$^{-1}$.

## 4. Conclusions

The results presented in the published papers [42, 53–63] show that the theoretical computations are an extremely powerful tool in the research of uranium-containing compounds. Once the proper relativistic norm-conserving pseudopotential has been generated [42, 52], the structural information, the X-ray powder patterns, the vibrational Raman spectra and mechanical and thermodynamic properties of these substances can be determined. This result is very significant because, if we have access to the adequate computational resources, very accurate results can be obtained despite the large size of the systems and the fact that the level of theory required to describe them is very high [64, 65]. The use of these methods is free of the difficulties of the experimental methods associated to the radiotoxicity of these compounds requiring a careful management of the samples. Thus, the theoretical calculations allow the safe study of secondary phases of spent nuclear fuel in definitive disposal conditions. The natural samples used in the experimental studies are generally mixtures of several minerals and the theoretical treatment allows to study pure substances. Furthermore, the synthesis of these compounds is very complex and generally produces samples with low crystallinity.

The theoretical solid-state methods may be used, in conjunction with experimental techniques, as an interpretative tool of the experimental structural and vibrational data or as a predictive tool to determine the structural, vibrational, mechanical and thermodynamic properties of these substances. The understanding of the structures of these compounds is very important itself to characterize them and to evaluate the possible incorporation of transuranic elements and fission products into the structures of uranyl minerals [8, 26–41]. The assignment of the main bands in the vibrational spectra of these compounds, performed usually by the experimentalist in an empirical way, can be made in a rigorous form since the theoretical methods produce microscopic scale views of the motion of the atoms in the corresponding normal vibrational states. In the systems studied the theoretical calculations has permitted the correct assignment of the bands of the full Raman spectrum for the first time. The main bands used to fingerprint these minerals were put into correspondence with specific structural data.

The calculated mechanical properties obtained for rutherfordine, soddyite and uranophane and schoepite minerals [53–56] have demonstrated the mechanical stability of their structures. Besides, a large amount of relevant mechanical data of these minerals were reported, including bulk modulus and its derivatives, elastic coefficients, shear and Young moduli, Poisson ratios, ductility and hardness indices, and elastic anisotropy measures. Their equations of state where also determined by fitting lattice volumes and pressures to a fourth order Birch-Murnaghan equation of state. The importance of the availability of these mechanical data cannot be overlooked. The large volume expansion of the SNF resulting from SNF corrosion during storage [127, 128] will cause a very large stress upon the waste matrix and therefore the mechanical behavior of the waste components is extremely relevant.

## Acknowledgements

This work has been carried out in the context of a CSIC-CIEMAT collaboration agreement: "Caracterización experimental y teórica de fases secundarias y óxidos de uranio formados en condiciones de almacenamiento de combustible nuclear". I also want to thank Dr. Rafael Escribano, Dr. Ana M. Fernández, Dr. Vicente Timón, Dr. Laura J. Bonales and Dr. Joaquín Cobos for their continuous help and advice during the realization of these studies.

## Conflict of interest

The author declares that there is no conflict of interest.

## Dedication

To my brother, TCR, on the occasion of his recent birthday.

## Permissions

Rutherfordine mineral. **Figures 1A**, **B** and **3A** and data contained in **Table 1** reproduced from Ref. [42] with permission from the PCCP Owner Societies. Data contained in **Tables 2–4** reprinted (adapted) with permission from Ref. [53], Copyright (2017) American Chemical Society.

Uranophane-$\alpha$ mineral. **Figures 1C**, **D** and **3C** and data contained in **Tables 1–4** reproduced with the permission of the Mineralogical Society of Great Britain & Ireland, from Ref. [55].

Soddyite mineral. **Figures 2A**, **B**, **3B**, **5–7** and data contained in **Tables 1–5** reprinted (adapted) with permission from Ref. [54], Copyright (2017) Elsevier.

Schoepite mineral. **Figures 2C**, **D**, **3D** and **4** and data contained in **Tables 1–4** reprinted (adapted) with permission from Ref. [56], Copyright (2018) American Chemical Society.

## Author details

Francisco Colmenero Ruiz

Address all correspondence to: francisco.colmenero@iem.cfmac.csic.es

Instituto de Estructura de la Materia – Consejo Superior de Investigaciones Científicas (IEM-CSIC), Madrid, Spain

## References

[1] Bruno J, Ewing RC. Spent nuclear fuel. Elements. 2006;**2**:343-349. DOI: 10.2113/gselements.2.6.343

[2] Ewing RC. The nuclear fuel cycle: A role for mineralogy and geochemistry. Elements. 2006;**2**:331-334. DOI: 10.2113/gselements.2.6.331

[3] Ewing RC. Long-term storage of spent nuclear fuel. Nature Materials. 2005;**14**:252-256. DOI: 10.1038/nmat4226

[4] Ewing RC, Weber WJ, Lian J. Nuclear waste disposal—Pyrochlore ($A_2B_2O_7$): Nuclear waste form for the immobilization of plutonium and "minor" actinides. Journal of Applied Physics. 2004;**95**:5949-5971. DOI: 10.1063/1.1707213

[5] Burns PC, Ewing RC, Navrotsky A. Nuclear fuel in a reactor accident. Science. 2012; **335**:1184-1188. DOI: 10.1126/science.1211285

[6] Geipel G, Thieme M. Determination of inorganic species in seepage water of uranium mining rockpiles and in related media. Journal of Radioanalytical Inorganic Nuclear Chemistry. 1984;**183**:139-145 DOI: 0.1007/BF02043127

[7] Abdelouas A. Uranium mill tailings: Geochemistry, mineralogy, and environmental impact. Elements. 2006;**2**:335-342. DOI: 10.2113/gselements.2.6.335

[8] Burns PC, Klingensmith AL. Uranium mineralogy and neptunium mobility. Elements. 2006;**2**:351-356. DOI: 10.2113/gselements.2.6.351

[9] Swedish Nuclear Fuel and Waste Management Co., SKB. Final Disposal of Spent Nuclear Fuel. Importance of the Bedrock for Safety. SKB Report 92-20; Stockholm: SKB; 1992

[10] Christensen H, Sunder S. An evaluation of layer water thickness effective in the oxidation of $UO_2$ fuel due to radiolysis of water. Journal of Nuclear Materials. 1996;**238**:70-77. DOI: 10.1016/S0022-3115(96)00342-X

[11] Wang R, Katayama YB. Dissolution mechanisms for $UO_2$ and spent fuel. Nuclear and Chemical Waste Management. 1982;**3**:83-90. DOI: 10.1016/0191-815X(82)90054-7

[12] Sattonnay G, Ardois C, Corbel C, Lucchini JF, Barthe MF, Garrido F, Gosset D. Alpha-radiolysis effects on $UO_2$ alteration in water. Journal of Nuclear Materials. 2001;**288**:11-19. DOI: 10.1016/S0022-3115(00)00714-5

[13] Shoesmith DW. Fuel corrosion processes under waste disposal conditions. Journal of Nuclear Materials. 2000;**282**:1-31. DOI: 10.1016/S0022-3115(00)00392-5

[14] Sunder S. Calculation of radiation dose rates in a water layer in contact with used CANDU $UO_2$ fuel. Nuclear Technology. 1998;**122**:211-221. DOI: 10.13182/NT98-A2863

[15] Roth O, Jonsson M. Oxidation of $UO_2$ (s) in aqueous solution. Central European Journal of Chemistry. 2008;**6**:1-14. DOI: 0.2478/s11532-007-0067-z

[16] Plasil J. Oxidation-hydration weathering of uraninite: The current state-of knowledge. Journal of Geosciences. 2014;**59**:99-114. DOI: 10.3190/jgeosci.163

[17] Grenthe I, Drozdzynski J, Fujino T, Buck EC, Albrecht-Schmitt TE, Wolf SF. Uranium. In: Morss LR, Edelstein NM, Fuger J, editors. The Chemisty of Actinide and Transactinide Elements. Vol. 1. Berlin: Springer Science and Business Media; 2006. Chapter V. pp. 253-638. ISBN: 978-94-007-0211-0

[18] Finch RJ, Ewing RC. The corrosion of uraninite under oxidizing conditions. Journal of Nuclear Materials. 1992;**190**:133-156. DOI: 10.1016/0022-3115(92)90083-W

[19] Frondel C. Mineral composition of gummite. American Mineralogist. 1956;**41**:539-568. ISSN: 0003-004X

[20] Frondel C. Systematic mineralogy of uranium and thorium. U.S. Geological Survey Bulletin. 1958;**1064**:1-400. Available from: http://pubs.er.usgs.gov/publication/b1064ER [Accessed: 15-05-2018]

[21] Pearcy EC, Prikryl JD, Murphy WM, Leslie BW. Alteration of uraninite from the Nopal I deposit, Peña Blanca district, Chihuahua, Mexico, compared to degradation of spent

nuclear fuel in the proposed US high-level nuclear waste repository at Yucca Mountain, Nevada. Applied Geochemistry. 1994;**9**:713-732. DOI: 10.1016/0883-2927(94)90030-2

[22] Wronkiewicz DJ, Bates JK, Gerding TJ, Veleckis E, Tani BS. Uranium release and secondary phase formation during unsaturated testing of $UO_2$ at 90°C. Journal of Nuclear Materials. 1992;**190**:107-127. DOI: 10.1016/0022-3115(92)90081-U

[23] Wronkiewicz DJ, Bates JK, Wolf SF, Buck EC. Ten-year results from unsaturated drip tests with $UO_2$ at 90°C: Implications for the corrosion of spent nuclear fuel. Journal of Nuclear Materials. 1996;**238**:78-95. DOI: 10.1016/S0022-3115(96)00383-2

[24] Finch RJ, Murakami T. Systematics and paragenesis of uranium minerals. In: Ribbe PH, editor. URANIUM: Mineralogy, Geochemistry and the Environment; Reviews in Mineralogy. Vol. 38. New York: Mineralogical Society of America; 1999. pp. 91-180. ISBN: 0-939950-50-2

[25] Clark DL, Hobart DE, Neu MP. Actinide carbonte complexes and their importance in actinide environmental chemistry. Chemical Reviews. 1995;**95**:25-48. DOI: 10.1021/cr00033a002

[26] Burns PC, Ewing RC, Miller ML. Incorporation mechanisms of actinide elements in to the structures of $U^{6+}$ phases formed during the oxidation of spent nuclear fuel. Journal of Nuclear Materials. 1997;**245**:1-9. DOI: 10.1016/S0022-3115(97)00006-8

[27] Burns PC. The crystal chemistry of uranium. In: Ribbe PH, editor. Uranium: Mineralogy, Geochemistry and the Environment; Reviews in Mineralogy. New York: Mineralogical Society of America; 1999. pp. 23-90. ISBN: 0-939950-50-2

[28] Burns PC. $U^{6+}$ minerals and inorganic compounds: Insights into an expanded structural hierarchy of crystal structures. The Canadian Mineralogist. 2005;**43**:1839-1894. DOI: 10.2113/gscanmin.43.6.1839

[29] Burns PC. The structure of Botwoodite and implications of solid solution towards sodium Boltwoodite. The Canadian Mineralogist. 1998;**36**:1069-1075. ISSN: 0008-4476

[30] Burns PC. Cs Boltwoodite obtained by ion exchange from single crystals: Implications for radionuclide release in a nuclear repository. Journal of Nuclear Materials. 1999;**265**:218-223. DOI: 10.1016/S0022-3115(98)00646-1

[31] Burns PC, Li Y. The structures of becquerelite and Sr-exchanged becquerelite. American Mineralogist. 2002;**87**:550-557. DOI: 10.2138/am-2002-0418

[32] Klingensmith AL, Burns PC. Neptunium substitution in synthetic uranophane and soddyite. American Mineralogist. 2007;**92**:1946-1951. DOI: 10.2138/am.2007.2542

[33] Burns PC, Deely KM, Skanthakumar S. Neptunium incorporation into uranyl compounds that form as alteration products of spent nuclear fuel: implications for geologic repository performance. Radiochimica Acta. 2004;**92**:151-159. DOI: 10.1524/ract.92.3.151.30491

[34] Burns PC, Miller ML, Ewing RC. $U^{6+}$ minerals and inorganic phases: A comparison and hierarchy of crystal structures. The Canadian Mineralogist. 1996;**34**:845-880. ISSN: 0008-4476

[35] Burns PC, Ewing RC, Hawthorne FC. The crystal chemistry of hexavalent uranium: Polyhedron geometries, bond-valence parameters, and polymerization of polyhedra. The Canadian Mineralogist. 1997;**35**:1551-1570. ISSN: 0008-4476

[36] Douglas M, Clark SB, Utsunomiya S, Ewing RC. Cesium and strontium incorporation into uranophane, $Ca[(UO_2)(SiO_3OH)]_2 \cdot 5H_2O$. Journal of Nuclear Science and Technology Supplement. 2002;**3**:504-507. DOI: 10.1080/00223131.2002.10875517

[37] Douglas M, Clark SB, Friese JI, Arey BW, Buck EC, Hanson BD, Utsunomiya S, Ewing RC. Microscale characterization of uranium(VI) silicate solids and associated neptunium(V). Radiochimica Acta. 2005;**93**:265-272. DOI: 10.1524/ract.93.5.265.64281

[38] Murphy WM, Grambow B. Thermodynamic interpretation of neptunium coprecipitation in uranophane for application to the Yucca Mountain Repository. Radiochimica Acta. 2008;**96**:563-567. DOI: 10.1524/ract.2008.1537

[39] Shuller LC, Ewing RC, Becker U. Quantum-mechanical evaluation of Np-incorporation into studtite. American Mineralogist. 2010;**95**:1151-1160. ISSN: 0003-004X

[40] Shuller LC, Ewing RC, Becker U. Np-incorporation into uranyl phases: A quantum–mechanical evaluation. Journal of Nuclear Materials. 2013;**434**:440-450. DOI: 10.1016/j.jnucmat.2011.04.016

[41] Shuller LC, Bender WM, Walker SM, Becker U. Quantum-mechanical methods for quantifying incorporation of contaminants in proximal minerals. Minerals. 2014;**4**:690-715. DOI: 10.3390/min4030690

[42] Bonales LJ, Colmenero F, Cobos J, Timón V. Spectroscopic Raman characterization of rutherfordine: A combined DFT and experimental study. Physical Chemistry Chemical Physics. 2016;**18**:16575-16584. DOI: 10.1039/C6CP01510G

[43] Baker RJ. Uranium minerals and their relevance to long term storage of nuclear fuels. Coordination Chemistry Reviews. 2014;**266-267**:123-136. DOI: 10.1016/j.ccr.2013.10.004

[44] Dubessy J, Caumon MC, Rull F, editors. Raman Spectroscopy Applied to Earth Sciences and Cultural Heritage. EMU Notes in Mineralogy. Vol. 12. European Mineralogical Union; 2012. ISBN: 978-0-903056-34-2

[45] Frost RL, Čejka J. A Raman spectroscopic study of the uranyl carbonate rutherfordine. Journal of Raman Spectroscopy. 2007;**38**:1488-1493. DOI: 10.1002/jrs.1796

[46] Frost RL, Čejka J. A Raman spectroscopic study of the uranyl mineral rutherfordine—Revisited. Journal of Raman Spectroscopy. 2009;**40**:1096-1103. DOI: 10.1002/jrs.2174

[47] Frost RL, Čejka J, Weier ML, Martens W. Molecular structure of the uranyl silicates—A Raman spectroscopic study. Journal of Raman Spectroscopy. 2006;**37**:538-551. DOI: 10.1002/jrs.1430

[48] Frost RL, Čejka J, Weier ML, Martens W. A Raman and infrared spectroscopic study of the uranyl silicates—Weeksite, soddyite and haiweeite: Part 2. Spectrochimica Acta A. 2006;**63**:305-312. DOI: 10.1016/j.saa.2005.05.015

[49] Frost RL, Čejka J, Weier ML. Raman spectroscopic study of the uranyl oxyhydroxide hydrates: becquerelite, billietite, curite, schoepite and vandendriesscheite. Journal of Raman Spectroscopy. 2007;**38**:460-466. DOI: 10.1002/jrs.1669

[50] Frost RL, Dickfos MJ, Čejka J. Raman spectroscopic study of the uranyl carbonate mineral zellerite. Journal of Raman Spectroscopy. 2008;**39**:582-586 DOI: 0.1002/jrs.1879

[51] Frost RL, Čejka J, Scholz R, López A, Theiss FL, Xi Y. Vibrational spectroscopic study of the uranyl selenite mineral derriksite $Cu_4UO_2(SeO_3)_2(OH)_6{\cdot}H_2O$. Spectrochimica Acta A. 2014;**113**:473-477. DOI: 10.1016/j.saa.2013.08.026

[52] Colmenero F. Characterization of secondary phases of spent nuclear fuel under final geological disposal conditions [Ph.D. thesis]. Madrid: Universidad Autónoma de Madrid; 2017. 443 pp. DOI: 10.13140/RG.2.2.10526.43843

[53] Colmenero F, Bonales LJ, Cobos J, Timón V. Thermodynamic and mechanical properties of rutherfordine mineral based on density functional theory. Journal of Physical Chemistry C. 2017;**121**:5994-6001. DOI: 10.1021/acs.jpcc.7b00699

[54] Colmenero F, Bonales LJ, Cobos J, Timón V. Structural, mechanical and vibrational study of uranyl silicate mineral soddyite by DFT calculations. Journal of Solid State Chemistry. 2017;**253**:249-257. DOI: 10.1016/j.jssc.2017.06.002

[55] Colmenero F, Bonales LJ, Timón V, Cobos J. Structural, mechanical and Raman spectroscopic characterization of layered uranyl silicate mineral uranophane-$\alpha$ by DFT Methods. Clay Minerals. 2018. DOI: 10.1180/clm.2018.27

[56] Colmenero F, Cobos J, Timón V. Periodic DFT study of the structure, Raman spectrum and mechanical properties of schoepite mineral. Inorganic Chemistry. 2018;**212**:125-135. DOI: 10.1021/acs.inorgchem.8b00150

[57] Colmenero F, Bonales LJ, Cobos J, Timón V. Study of the thermal stability of studtite by in situ Raman spectroscopy and DFT calculations. Spectrochimica Acta A. 2017;**174**:245-253. DOI: 10.1016/j.saa.2016.11.040

[58] Colmenero F, Bonales LJ, Cobos J, Timón V. Density functional theory study of the thermodynamic and Raman vibrational properties of $\gamma$-$UO_3$ polymorph. Journal of Physical Chemistry C. 2017;**121**:14507-14516. DOI: 10.1021/acs.jpcc.7b04389

[59] Colmenero F, Fernández AM, Cobos J, Timón V. Thermodynamic properties of uranyl-containing materials based on density functional theory. Journal of Physical Chemistry C. 2018;**122**:5254-5267. DOI: 10.1021/acs.jpcc.7b12341

[60] Colmenero F. The application of periodic density functional theory to the study of uranyl containing materials: Thermodynamic properties and stability. In: Glossman-Mitnik D, editor. Density Functional Theory. London: InTechOpen; 2018. ISBN 978-953-51-7020-4

[61] Colmenero F, Fernández AM, Cobos J, Timón V. Temperature-dependent gibbs free energies of reaction of uranyl-containing materials based on density functional theory. Journal of Physical Chemistry C. 2018;**122**:5268-5279. DOI: 10.1021/acs.jpcc.7b12368

[62] Colmenero F, Fernández AM, Cobos J, Timón V. Periodic DFT study of the thermodynamic properties and stability of schoepite and metaschoepite mineral phases. Physical Chemistry Chemical Physics. 2018 (under review)

[63] Colmenero F, Fernández AM, Cobos J, Timón V. Becquerelite mineral phase: crystal structure and thermodynamic and mechanic stability by using periodic DFT. RSC Advances. 2018;**8**:24599-24616. DOI: 10.1039/c8ra04678f

[64] Crocombette JP, Jollet F, Nga LT, Petit T. Plane-wave pseudopotential study of point defects in uranium dioxide. Physical Review B. 2001;**64**:104107. DOI: 10.1103/PhysRevB.64.104107

[65] Beridze G, Kowalski PM. Benchmarking the DFT+U method for thermochemical calculations of uranium molecular compounds and solids. Journal of Physical Chemistry A. 2014;**118**:11797-11810. DOI: 10.1021/jp5101126

[66] Burns PC. CCD area detectors of X-rays applied to the analysis of mineral structures. The Canadian Mineralogist. 1998;**36**:847-853. ISSN: 0008-4476

[67] Payne MC, Teter MP, Ailan DC, Arias A, Joannopoulos JD. Iterative minimization techniques for Ab initio total-energy calculations: Molecular dynamics and conjugate gradients. Review of Modern Physics. 1992;**64**:1045-1097. DOI: 10.1103/RevModPhys.64.1045

[68] Weck PF, Kim E, Buck EC. On the mechanical stability of uranyl peroxide hydrates: Implications for nuclear fuel degradation. RSC Advances. 2015;**5**:79090-79097. DOI: 10.1039/C5RA16111H

[69] Weck PF, Kim E. Uncloaking the thermodynamics of the studtite to metastudtite shear-induced transformation. Journal of Physical Chemistry C. 2016;**120**:16553-16560. DOI: 10.1021/acs.jpcc.6b05967

[70] Fritz I. Elastic properties of $UO_2$ at high pressure. Journal of Applied Physics. 1976;**47**:4353. DOI: 10.1063/1.322438

[71] Ibrahim M, Castelier E, Palancher H, Bornert M, Care S, Micha JS. Mechanical behaviour near grain boundaries of He-implanted $UO_2$ ceramic polycrystals. Journal of Nuclear Materials. 2017;**483**:13-20. DOI: 10.1016/j.jnucmat.2016.10.044

[72] Nguyen TH, Debelle A, Boulle A, Garrido F, Thomas L, Demange V. Mechanical response of $UO_2$ single crystals submitted to low-energy ion irradiation. Journal of Nuclear Materials. 2015;**467**:505-511. DOI: 10.1016/j.jnucmat.2015. 10.046

[73] Nye JF. The Physical Properties of Crystals: Their Representation by Tensors and Matrices. New York: Oxford University Press; 1985. ISBN: 0198511655

[74] Baroni S, de Gironcoli S, Dal Corso SA. Phonons and related crystal properties from density-functional perturbation theory. Review of Modern Physics. 2001;**73**:515-562. DOI: 10.1103/RevModPhys.73.515

[75] Gonze X, Lee C. Dynamical matrices, born effective charges, dielectric permittivity tensors, and interatomic force constants from density-functional perturbation theory. Physical Review B. 1997;**55**:10355-10368. DOI: 10.1103/PhysRevB.55.10355

[76] Refson K, Tulip PR, Clark SJ. Variational density-functional perturbation theory for dielectrics and lattice dynamics. Physical Review B. 2006;**73**:155114. DOI: 10.1103/PhysRevB.73.155114

[77] Yun Y, Legut D, Oppeneer PM. Phonon spectrum, thermal expansion and heat capacity of $UO_2$ from first-principles. Journal of Nuclear Materials. 2012;**426**:109-114. DOI: 10.1016/j.jnucmat.2012.03.017

[78] Yin Q, Savrasov SY. Origin of low thermal conductivity in nuclear fuels. Physical Review Letters. 2008;**100**:225504. DOI: 10.1103/PhysRevLett.100.225504

[79] Wang BT, Zhang P, Lizárraga P, Di Marco I, Eriksson O. Phonon spectrum, thermodynamic properties, and pressure-temperature phase diagram of uranium dioxide. Physical Review B. 2013;**88**:104107. DOI: 10.1103/PhysRevB.88.104107

[80] Mei ZG, Stan M, Yang J. First-principles study of thermophysical properties of uranium dioxide. Journal of Alloys and Compounds. 2014;**603**:282-286. DOI: 10.1016/j.jallcom.2014.03.091

[81] Mei ZG, Stan M, Pichler B. First-principles study of structural, elastic, electronic, vibrational and thermodynamic properties of UN. Journal of Nuclear Materials. 2013;**440**:63-69. DOI: 10.1016/j.jnucmat.2013.04.058

[82] Perdew JP, Burke K, Ernzerhof M. Generalized gradient approximation made simple. Physical Review Letters. 1996;**77**:3865-3868. DOI: 10.1103/PhysRevLett.77.3865

[83] Grimme S. Semiempirical GGA-type density functional constructed with a long-range dispersion correction. Journal of Computational Chemistry. 2006;**27**:1787-1799. DOI: 10.1002/jcc.20495

[84] Perdew JP, Ruzsinszky A, Csonka GI, Vydrov OA, Scuseria GE, Constantin LA, Zhou X, Burke K. Restoring the density-gradient expansion for exchange in solids and surfaces. Physical Review Letters. 2008;**100**:136406. DOI: 10.1103/PhysRevLett.100.136406

[85] Clark SJ, Segall MD, Pickard CJ, Hasnip PJ, Probert MIJ, Refson K, Payne MC. First principles methods using CASTEP. Zeitschrift für Kristallographie-Crystalline Materials. 2005;**220**:567-570. DOI: 10.1524/zkri.220.5.567.65075

[86] Materials Studio. Available from: http://accelrys.com/products/collaborativescience/biovia-materials-studio/ [Accessed: 30-04-2018]

[87] Troullier N, Martins JL. Efficient pseudopotentials for plane-wave calculations. Physical Review B. 1991;**43**:1993-2006. DOI: 10.1103/PhysRevB.43.1993

[88] Pfrommer BG, Cote M, Louie MSG, Cohen ML. Relaxation of crystals with the Quasi-Newton method. Journal of Computational Physics. 1997;**131**:233-240. DOI: 10.1006/jcph.1996.5612

[89] Monkhorst HJ, Pack JD. Special points for Brillonin-zone integration. Physical Review B. 1976;**13**:5188-5192. DOI: 10.1103/PhysRevB.13.5188

[90] Yu R, Zhu RJ, Ye HQ. Calculations of single-crystal elastic constants made simple. Computer Physics Communications. 2010;**181**:671-675. DOI: 10.1016/j.cpc.2009.11.017

[91] Birch F. Finite elastic strain of cubic crystal. Physical Review. 1947;**71**:809-824. DOI: 10.1103/PhysRev.71.809

[92] Angel RJ. Equations of state. Reviews in Mineralogy and Geochemistry. 2000;**41**:35-60. DOI: 10.2138/rmg.2000.41.2. EOSFIT 5.2 software, http://www.ccp14.ac.uk/ccp/webmirrors/ross-angel/rja/soft/.10.2138/rmg.2000.41.2

[93] Colmenero F, Timon V. Study of the structural, vibrational and thermodynamic properties of natroxalate mineral using density functional theory. Journal of Solid State Chemistry. 2018;**263**:131-140. DOI: 10.1016/j.jssc.2018.04.022

[94] Colmenero F. Structural, spectroscopic and thermodynamic characterization of ammonium oxalate monohydrate mineral by means of theoretical solid-state methods. Journal of Physics and Chemistry of Solids. 2018. (accepted for publication)

[95] Milman V, Refson K, Clark SJ, Pickard CJ, Yates JR, Gao SP, Hasnip PJ, Probert MIJ, Perlov A, Segall MD. Electron and vibrational spectroscopies using DFT, plane waves and pseudopotentials: CASTEP implementation. Journal of Molecular Structure THEOCHEM. 2010;**954**:22-35. DOI: 10.1016/j.theochem.2009.12.040

[96] Hehre WJ, Radom L, Schleyer PVR, Pople JA. Ab Initio Molecular Orbital Theory. New York: Wiley; 1986. ISBN: 0471812412

[97] Finch RJ, Cooper MA, Hawthorne FC, Ewing RC. Refinement of the crystal structure of rutherfordine. The Canadian Mineralogist. 1999;**37**:929-938. ISSN: 0008-4476

[98] Ginderow D. Structure de l'Uranophane Alpha, $Ca(UO_2)_2(SiO_3OH)_2.5H_2O$. Acta Crystallographica C. 1988;**44**:421-424. DOI: 10.1107/S0108270187011491

[99] Demartin F, Gramaccioli CM, Pilati T. The importance of accurate crystal structure determination of uranium minerals. II. Soddyite $(UO_2)_2(SiO_4)\cdot 2H_2O$. Acta Crystallographica C. 1992;**48**:1-4. DOI: 10.1107/S0108270191004481

[100] Finch RJ, Cooper MA, Hawthorne FC, Ewing RC. The crystal structure of schoepite, $[(UO_2)_8 O_2(OH)_{12}]\cdot(H_2O)_{12}$. The Canadian Mineralogist. 1996;**34**:1071-1088. ISSN: 0008-4476

[101] Christ CL, Clark JR. Some observations on rutherfordine. American Mineralogist. 1956; **41**:844-850. ISSN: 0003-004X

[102] Christ CL, Clark JR, Evans T. Crystal structure of rutherfordine, $UO_2CO_3$. Science. 1955;**121**:472-473. DOI: 10.1126/science.121.3144.472

[103] Stohl FV, Smith DK. The crystal chemistry of the uranyl silicate minerals. American Mineralogist. 1981;**66**:610-624. ISSN: 0003-004X

[104] Evans HT. Uranyl ion coordination. Science. 1963;**141**:154-158. DOI: 10.1126/science.141.3576.154

[105] Weller MT, Light ME, Gelbrich T. Structure of uranium(VI) oxide dihydrate, $UO_3\cdot 2H_2O$; synthetic meta-schoepite $(UO_2)_4O(OH)_6\cdot 5H_2O$. Acta Crystallographica B. 2000;**56**:577-583. DOI: 10.1107/S0108768199016559

[106] Piret P. Structure cristalline de la fourmarierite, $Pb[(UO_2)_4O_3(OH)_4]·4H_2O$. Bulletin de la Société Française de Minéralogie et de Cristallographie. 1985;**108**:659-665. ISSN: 0037-9328

[107] Downs RT, Bartelmehs KL, Gibbs GV, Boisen MB. Interactive software for calculating and displaying X-ray or neutron powder diffractometer patterns of crystalline materials. American Mineralogist. 1993;**78**:1104-1107. ISSN: 0003-004X

[108] Martin JD. XPowder12. Ver. 04.13, 2012. Available from: http://www.xpowder.com/download/xpowder.pdf [Accessed: 30-04-2018]

[109] ICDD. The International Center for Diffraction Data. PDF-2 Database 2003. Available from: http://www.icdd.com/index.php/pdf-2/ [Accessed: 30-04-2018]

[110] Downs RT. The RRUFF Project: An Integrated Study of the Chemistry, Crystallography, Raman and Infrared Spectroscopy of Minerals. Program and Abstracts of the 19th General Meeting of the International Mineralogical Association in Kobe. Japan; 2006. O03-13. Available from: RRUFF Database web page, http://rruff.info/schoepite [Accessed: 30-04-2018]

[111] Born M. On the stability of crystal lattices. I. Mathematical Proceedings of the Cambridge Philosophical Society. 1940;**36**:160-172. DOI: 10.1017/S0305004100017138

[112] Born M, Huang K. Dynamics Theory of Crystal Lattices. Oxford: Oxford University Press; 1954. ISBN:0-19-850369-5

[113] Mouhat F, Coudert FX. Necessary and sufficient elastic stability conditions in various crystal systems. Physical Review B. 2014;**90**:224104. DOI: 10.1103/PhysRevB.90.224104

[114] Voigt W. Lehrbuch der Kristallphysik mit Ausschluss der Kristalloptik. Leipzig: Teubner; 1962. ISBN: 3663153169

[115] Reuss A. Berechnung der Fliessgrenze von Mischkristallen auf Grund der Plastizitatsbedingung fur Einkristalle. Journal of Applied Mathematics and Mechanics. 1929;**9**:49-58. DOI: 10.1002/zamm.19290090104

[116] Hill R. The elastic behaviour of a crystalline aggregate. Proceedings of the Physical Society of London. 1952;**65**:349-354. DOI: 10.1088/0370-1298/65/5/307

[117] Bouhadda Y, Djella S, Bououdina M, Fenineche N, Boudouma Y. Structural and elastic properties of $LiBH_4$ for hydrogen storage applications. Journal of Alloys and Compounds. 2012;**534**:20-24. DOI: 10.1016/j.jallcom.2012.04.060

[118] Ulian G, Tosoni S, Valdre G. The compressional behaviour and the mechanical properties of talc $[Mg_3Si_4O_{10}(OH)_2]$: A density functional theory investigation. Physics and Chemistry of Minerals. 2014;**41**:639-650. DOI: 10.1007/s00269-014-0677-x

[119] Shang S, Wang Y, Liu ZK. First-principles elastic constants of $\alpha$-and $\theta$-$Al_2O_3$. Applied Physics Letters. 2007;**90**:101909. DOI: 10.1063/1.2711762

[120] Marmier A, Lethbridge ZAD, Walton RI, Smith CW, Parker SC, Evans KE. ElAM: A computer program for the analysis and representation of anisotropic elastic properties. Computer Physics Communications. 2010;**181**:2102-2115. DOI: 10.1016/j.cpc.2010.08.033

[121] Ravindran P, Fast L, Korzhavyi PA, Johansson B, Wills J, Eriksson O. Density functional theory for calculation of elastic properties of orthorhombic crystals: Application to $TiSi_2$. Journal of Applied Physics. 1998;**84**:4891-4904. DOI: 10.1063/1.368733

[122] Pugh SF XCII. Relations between the elastic moduli and the plastic properties of polycrystalline pure metals. Philosophical Magazine. 1954;**45**:823-843. DOI: 10.1080/14786440808520496

[123] Niu H, Wei P, Sun Y, Chen CX, Franchini C, Li D, Li Y. Electronic, optical, and mechanical properties of superhard cold-compressed phases of carbon. Applied Physics Letters. 2011;**99**:031901. DOI: 10.1063/1.3610996

[124] Ranganathan SI, Ostoja-Starzewski M. Universal elastic anisotropy index. Physical Review Letters. 2008;**101**:055504. DOI: 10.1103/PhysRevLett.101.055504

[125] Wheaton V, Majumdar D, Balasubramanian K, Chauffe L, Allen PGA. Comparative theoretical study of uranyl silicate complexes. Chemical Physics Letters. 2003;**371**:349-359. DOI: 10.1016/S0009-2614(03)00237-9

[126] Biwer DM, Ebert WL, Bates JK. The Raman spectra of several uranyl-containing minerals using a microprobe. Journal of Nuclear Materials. 1990;**175**:188-193. DOI: 10.1016/0022-3115(90)90205-2

[127] Schmidt AJ, Delegard CH. Assessment of K Basin Sludge Volume Expansion Resulting from Uranium Corrosion During Storage. Report of the U.S. Department of Energy, Pacific Northwest National laboratory PNNL-13786. Richland, Washington: US-DOE; 2002

[128] Delegard CH, Schmidt AJ, Chenault, JW. Mechanical Properties of K Basin Sludge Constituents and Their Surrogates. Report of the U.S. Department of Energy, Pacific Northwest National laboratory PNNL-14947. Richland, Washington: US-DOE; 2004

Chapter 5

# Distribution Law for Constituent Minerals and Chemical Components in Rocks and Ores

Basanta K. Sahu

Additional information is available at the end of the chapter

http://dx.doi.org/10.5772/intechopen.88125

**Abstract**

Rocks and Ores comprising of multicomponent of minerals (Group I) contain chemical molecules (Group II) which further contain (Group III). Man has exploited these valuable constituents for industrial, economic, and social growth causing serious depletion of high-grade ores at surface and shallow depths. Sustainable growth now requires exploitation of low-grade ores and those occurring at depth which imply optimal and most efficient mining efforts based on spatial, temporal, and spatiotemporal distribution of these constituents. The optimal decisions resulting in profit maximization is possible by obtaining precise and accurate parameters of these distributions. Sample size required for such estimations must be at least representative elementary volume (REV) of the rocks/ores, but the data matrix is not full-rank for statistical analyses and sample mean ($x$; $0 < x < 1$) must be transformed to Gaussian to apply standard univariate (UND)/multivariate (MND) statistical techniques. A $\log(x/(1-x))$ or $\ln(x/(1-x))$ transform is shown to be an appropriate pre-transformation that eliminates the twin problems of full-rank, and spurious negative correlations as well as makes the distribution Gaussian for major, minor, and trace components. Mining applications using the univariate and/or Multivariate Normal Theory of pre-transformed sample mean ($x$) in rocks/ores is optimal for anomaly detection, drilling site selection, global reserve and grade estimations, mine planning, ore mineral liberation, blending, sustainable mine developments and maximization of profits computed on net profit value (NPV) basis to present time.

**Keywords:** hierarchical geochemical constituents, fractional mineral and chemical constituents ($x$; $0 < x < 1.0$ or 100%) in rocks and ores, logarithmic pre-transformation of sample data using $\log(x/(1 - x))$, average Gaussian, statistical and geological inferences for reserves and average grades, mine life at minimum economic extractions, profit maximization through estimation of optimal cutoff grade

# 1. Introduction

Large-scale industrialization and urbanization with greater consumption of mineral and energy resources since the "Industrial Revolution" have greatly depleted high-grade resources available at shallow depths of crust [1–3]. Therefore, more intensive search for these finite resources is necessary at greater depths and for lower grades besides optimal mining, processing, and marketing in order to maintain sustainable development of economy and society and to have higher national growth. Mineral resources include high-grade marketable ores and/or associated low-grade ores which require blending and/or beneficiation to market the mined and/or processed products at profit [4, 5].

Rocks and ores comprise of multielement/multimineral materials containing more than three constituents ($C > 3$), some of which need separation and/or beneficiation to remove nonmarketable gangue minerals and waste materials [1, 6]. We are interested in static characterization (invariant in time) of the heterogeneous solids (rocks/ores) which can be achieved through sampling of adequate material (>representative elementary volume (REV) [7]) from the rock/ore, where REV means the minimum volume/weight that represents elementary volume. Thus, integration of any random variable (fractional constituent, x with $0 < x < 1$) over the REV provides stable and unbiased statistics representing the characteristics of an equivalent statistically isotropic homogeneous sample. The probability density functions (PDFs) corresponding to the random fractional constituents (such as minerals, molecules, elements, or isotopes) of the rock/ore are having a binomial distribution for major and minor constituents which reduces to a Poisson distribution for trace constituents. Realistic and accurate PDFs are essential for optimal exploration, mining, mine planning, processing, blending, beneficiation, and marketing of the concerned mineral resource. In contrast to rocks and ores, mineral geochemistry involves sampling of (homogeneous) minerals which would be much smaller in volume than REV for geochemical analysis and inference [8].

Major, minor, or trace fractional constituents ($x$; $0 < x < 1$) of rocks and ores are Gaussianized using the $\log(x)$ pre-transformation since 1954 as advocated by Ahrens [9]. However, this log (x) pre-transformation is not always valid for major constituents and not necessarily independent for these constituents; hence, a slight modification introducing multiplicative random errors and independence of transformed random constituents, such as $\log(x/(1 - x))$, would be useful for all three levels of constituents [10, 11]. Rocks and ores usually contain 5–6 minerals (some at major and minor and a few at trace levels) and about 10–12 major elements and a large number of trace elements, which can be explained by thermodynamic principles for closed systems having 2 degrees of freedom. However, open systems have a larger number of degrees of freedom and hence a larger number of phases (minerals).

From statistical viewpoint, the main problems of geochemical analysis and inference of such data (without pre-transformation) are:

- Constituents are usually weight, volume, area, or length fractions but not numbers as needed in statistics and probability.

- Weight fractions are not necessarily equal to volume, area, or length fractions (as given in stereology textbooks) since rocks/ores contain finite pore spaces.
- Statistics obtained in 3D space are not equal to corresponding statistics on length or number basis unless the total measure tends to be infinite (but sample size is finite).
- Probability distributions of constituents are not Gaussian for statistical hypotheses tests and inferences.

Four additional and more complex problems include:

i. Constant sum or closure: This constraint on fractional constituents induces spurious negative correlations among the constituents within samples which confound statistical and geochemical inference.

ii. Data matrix has a rank (C-1) and is not full rank (C) to have a unique inverse.

iii. Fractional constituents are not independent as required for parameter (such as mean and variance of order); two are finite and hence useful for parameter estimation for mean and variance, and all higher-order cumulants are zeros and hence dropped.

iv. Parameter estimation, hypotheses tests, and risk analyses are simple and straightforward on Gaussian data.

Mineral resources are non-replaceable national assets which should be optimally utilized for sustainable economic growth, social benefits, and improvements in health and life quality without jeopardizing present ecology/environment for future growth. Mineral resources are characteristically nonrenewable and form prime assets for stable economy and quality of life [2]. Optimal extraction and marketing strategies must be evolved under dynamic conditions of high risks in estimation of reserves and grades of marketable as well as lean-grade resources for optimizing the net profits at present value under highly fluctuating pricing and marketing conditions and also resource augmentation through exploration efforts and/or technological substitutions and proper waste disposals and of learning and updating. The main aim of the mining industry is to maximize the accrued profits using the concept of net present value (NPV) while simultaneously minimizing damage to ecology and environment, and improving the social and health needs of the local community seems essential to achieve conservation of high-grade ores and sustainable development. Capital must be generated at a faster rate than deployment for new discoveries or for R&D efforts for technological substitutions, and the accumulated capital thus helps in industrial and social growths and safeguards and improves life quality of people [1, 11–14].

Mineral resources not only comprise high-grade directly marketable ores but also contain considerable lean-grade nonmarketable ores as well as waste materials of little economic value which must be suitably treated and properly disposed to protect ecology/environment. Mineral deposits, including high- and/or lean-grade resources, are generally finite nonrenewable resources which will be exhausted under constant/varying rates of extraction in some finite time, called terminal time to exhaustion, T. This terminal time can be extended by augmentation by exploration, blending, and/or beneficiation of lean ores and technological substitutions but is still finite time. Here we consider resource augmentation through blending and/or beneficiation of lean-grade ores that cannot otherwise be marketed and create ecological hazards, and these techniques are much less risky than exploration efforts or R&D effort for substitutions. Hence, it would be optimal to extract resources with suitable treatments of blending and/or beneficiation of lean-grade ores that are inevitably associated with ore deposits. However, ore extraction rates and subsequent blending/beneficiation processes must be dynamically optimized, and the net present value of resource is maximized to provide a stable and viable industrial growth with benefits to all including mine owners, national and state governments, and public at large. Therefore, proper planning, extraction, and marketing policies would insure maximal national and social growths as well as sustainable life quality of people [2, 3, 5].

Mineral resources are considered to be national or state government properties and are usually allocated/leased to national and/or private parties for a fixed time period for extraction of ores. Ill effects of such allocations (such as many scams) can be overcome through various regulations, taxations, and royalty schemes. It is suggested that heavy penalty clauses must be included in these regulations so that parties involved would not try to break these laws. Market complexities do induce high fluctuations in future prices and costs and involve great risks in all mining operations. Risk-aversionist mine owners and mineral traders may plead for non optimal rapid depletion of valuables by nonrenewable high-grade ores. Hence, time paths generated by resource markets are not necessarily at equilibrium, and these paths are generally not the most efficient or optimal paths. Therefore, analyses of these past price and demand data must be performed with much care and foresight to provide sound decisions.

Policy measures are basic to operation of economic and market systems since they could reduce market volatility and increase accuracy of future expectations. Better forecast techniques such as ARIMA (p, d, q) for nonseasonal and SARIMA (P, D, Q) for seasonal time series data [10, 15] and widespread dissemination of these forecasts would be most useful to avert risks. Reduction of government secrecy about future intentions/policies and maintaining longer-term economic policies would yield better stability and benefits. When global price becomes low, mineral exporters should be adequately compensated for their loss and/or be provided with liberal credit facilities. If the terminal date of exhaustion, T, of a mine be known (or estimable with accuracy), then optimal extraction rate/depletion policy can be achieved for a closed economy with no import/export. Optimal rate of resource depletion depends on factors, such as discount rate, elasticity of marginal utility and substitute, and productivity of substitutes. But the generated model would still be extremely rough as accurate estimates of most of these technological parameters are not available.

Resource allocation for R&D activity to develop natural/technological substitute materials for scarce and exhaustible mineral resources is useful and very important. But, we know that finding a substitute is a random process where the date of invention is extremely uncertain (purely random) and may take as much as 20 years (lifetime of a mine). Hence, total cost of R&D may become very high, especially for small mines, and they should concentrate on small technological researches to improve mining, blending, beneficiation, marketing, etc. rather than to find technological substitutes. In any case, postponing intensive R&D activities to the last 5/6 years of a mine would be economically viable and prudent especially in a developing economy as India [14].

Although there exists extensive but finite low-grade resources which can be upgraded to marketable products by suitable blending and/or beneficiation techniques, this only extends life of mine which will finally be exhausted. The prime high-grade deposits are much less available and will be depleted in finite terminal time T (say, maximum of 70 years). Therefore, it would be prudent to estimate optimal rates of extraction, so that the net present value after discounting of the mine is maximized in the long run. Even if technological substitutes can be found in the future, the date of availability and amounts of product are random variables with unknown probability densities for any accurate estimation of their parameters [6, 16].

Optimal decisions are necessary at all stages of mining and marketing operations which are listed as follows:

a. **Mining stage:** Methodology adopted; mine plans in 3D based on the lowest mineable assay surfaces and associated risk factors; blocking of high-grade, lean-grade, and waste materials; dynamic extraction rates; horizontal and vertical extensions

b. **Blending and beneficiation:** Optimal lowest crushing and beneficiation sizes needed for blending and beneficiation [17]; identification of high-grade, lean-grade, and waste blocks for optimal mining and storing, etc.; conservation of high-grade ore blocks

c. **Waste management:** Wastes develop at mining, blending, and beneficiation stages; optimal treatment and safe disposal avoiding health hazards and damage to ecology

d. **Marketing:** Optimal classes of marketable ores by suitable mix of high-grade and lean-grade ores; future marketing of lean ores in situ or in dumps using blending and/or ore beneficiation

Thus, mining of nonrenewable resources (metallic, nonmetallic, industrial rocks, gems, etc.) are highly complex nonlinear risky dynamic processes and not yet completely understood or solved. The main problems are (i) how to allocate exhaustible resource so as to make its marginal social value equal for all uses and constant over time and (ii) how government should plan and conduct project exercises to achieve these goals. We highlight the utilization of nonmarketable lean ores by optimal blending and/or beneficiation techniques which would maximize economic growth, improve social benefits, and preserve ecology and environment (see Refs. [2, 8, 17–19] for more details).

## 2. Mathematical and statistical theory

### 2.1. Probability space and distribution for fractional constituents

Fractional constituents ($x$; $0 < x < 1$) belong to a C-dimensional positive real space and cannot be viewed geometrically if the number of constituents, C, is greater than three which is most frequent in rocks and ores. Aitchison [7] and his coworkers [20] have proposed that fractional constituents belong to a C-dimensional Simplex. They have used rather complex pre-transforms such as log ratio (lr), centered log ratio (clr), and isometric log ratio (ilr) to Gaussianize the PDFs for statistical analyses. However, Carranza [21] has shown that these complex pre-transformations still do not solve the rank problem (rank is C-1 instead of C) and inherent spurious negative correlations among the constituents within any sample. This is due to the fact that fractional constituents (x; $0 < x < 1$) belong to interior points of the simplex but not to apexes, hyper-edges, hyperplanes, and some sub-compositional hyperplanes of this C-dimensional simplex [6, 11].

From measure theory, fractional constituents (x; $0 < x < 1$) belong to the open interval (0, 1) and have a binomial/Poisson (not Gaussian) distribution excluding the non-admissible points 0 and 1 on this line. Since the fractional constituents are averaged values over the sampled REV and REV is much larger than the specific constituent, this random variable (rv) x can be considered to be a continuous rather than a discrete binomial/Poisson rv. It is absolutely necessary; otherwise, the rock/ore is not a C-dimensional simplex. In such a system, additive probability measure is not applicable, and a multiplicative probability model (odds ratio) leading to a log-Gaussian or lognormal model (as proposed by Ahrens in [9]) would be appropriate with some modification such that the log (odds ratios) of the constituents becomes independent and Gaussian (homogeneous variances).The simple modification for log (odds) as proposed in Sahu [10, 11, 16] is the following arguments:

If x is lognormal with mean ($\mu$) and variance ($\sigma^2$), then its complement $(1 - x)$ is also a fractional constituent. Hence, it would also follow lognormal distribution using theory of linear addition (or subtraction) of Gaussian random variables to be Gaussian (that is sums of Gaussian pdfs are closed/remain Gaussian). Therefore, we obtain $\log(x/(1 - x))$ to be Gaussian, and this pre-transformation involves only one constituent (x) of the rock/ore, which makes $\log(x/(1 - x))$ transform to be independent of all other constituents in the rock/ore [4, 10, 11, 16].

This $\log(x/(1 - x))$ pre-transformation Gaussianizes the pdfs of fractional concentrations while simultaneously eliminating the other four crucial drawbacks (i)–(iv) of binomial/Poisson distributions as listed earlier. Mathematical proof based on measure-theoretic analysis is given by Le Cam [22], Le Cam and Yang [23], and Sahu [11]. The author and his many students have used the $\log(x/(1 - x))$ pre-transformation to Gaussianize fractional concentrations (x) in geochemistry and apply to exploration, estimation, modeling, and hypothesis testing for several Indian ore deposits including those of gold, iron ores, lead-zinc, copper, phosphate, etc.

However, other log ratios such as lr, central-lr, and isometric-lr, proposed by Aitchison [7, 20], are not independent since the denominator constituent includes the numerator constituent as well, and, hence, the rank of data matrix still remains (C-1) or less (and not full rank of C, as is needed for its unique inverse) (see also [16, 21]).

### 2.2. Optimal extraction rates

#### *2.2.1. General considerations*

An exhaustible mineral deposit may be characterized by initial reserve, $S_0$, at present time, $t_0$, and rate of extraction, $R_t$, ($t > o$), at time t ($t>t_0$). The reserve at time t is $S_t$, given by

$$S_t = S_0 - \int_0^t R_t \, \mathrm{dt} \tag{1}$$

Thus, the definite integral of (Rt dt) from 0 to t is less than the initial reserve, So (2)

$$\text{Therefore, } \mathrm{d}S_t/\mathrm{dt} = -R_t, \text{ with } S_0 \text{ known and } S_t >= 0. \tag{3}$$

Eq. 2 gives in lim(t--> 0); $R_t$=0. However, $R_t$ cannot be predicted as present market price fluctuates over time and market price controls both economic extraction rates to keep profits positive and the terminal time to exhaustion, T, of mine. Assuming $R_t$ to be constant (R) over time T, then Eq. (1) gives

$$\mathrm{R} = S_0/\mathrm{T}, \text{ where } S_t = 0 \text{ (i.e., mine reached exhaustion).} \tag{4}$$

In mining and many other industries, usually time is discretized to mainly years but further into quarters, months, etc. Let price at time t be $p_t$ and rate of return/interest denoted as r. Price at time t + u,

$$p_{(t+u)} = \left(1 + r_t \, \mathrm{u} \, p_t\right) \text{ for } \mathrm{r(t)} > 0. \tag{5}$$

In limit u tending to 0, we have a support price of:

$$\left(\mathrm{d}p_t/\mathrm{dt}\right)/p_t = r_t \tag{6}$$

However, Eq. (6) does not include costs of mining, transportation, and marketing and, hence, is not useful in practice. We must maximize the net present value of property as.

$$P_t S_t = \mathrm{R}\left(\tau \ \exp.(-\mathrm{r}\,(\tau - \mathrm{t})\right) \mathrm{dt}, \tag{7}$$

where $P_t^R$ is spot price, $p_t$ is initial reserve stock, and R(τ) is extraction probability subject to the condition:

$$\int_t^\infty \{R(\tau) d(\tau)\} = S_t. \tag{8}$$

If R*(τ) is the optimal (maximal) solution to Eq. 6, then.

$$\left(\mathrm{d}p_t/\mathrm{dt}\right) S_t + p_t (\mathrm{d}S_t/\mathrm{dt}) = \mathrm{r}p_t S_t - P_t^R R_t{}^*). \tag{9}$$

From Eq. 7 we get $S_t$ = –R*, and hence we can write.

$$\left(\mathrm{d}p_t/\mathrm{dt} - \mathrm{r}p_t\right) = \left(p_t – P_t^R\right) \left(R_t{}^*/S_t\right). \tag{10}$$

Under competitive market stock price is equal to flow price (market equilibrium),

$p_t = P_t^R$, and Eq. (10) yields (d$p_t$/dt)/ $p_t$ = r which is the Hotelling Rule [24].

The price of mineral resource is expected to rise in the future with increasing population and consequent increase in demand as compared to the interest rate, r; the value of reserve (or stock) or maximum present value of sales is independent of the actual extraction policy provided the entire stock is exhausted over time. Thus $R_t$* is not time-invariant (constant) but varies with time t. Static and dynamic optimization of extraction rates, $R_t$*, are discussed in Sahu [2, 14], respectively.

### 2.3. Socially managed exhaustible resource

Let R = D(p) be the market demand decline curve for resource flow. Since dD(p)/dt is negative, we can invert to obtain p = $D^{-R}$= B(R) for convenience. The gross rate of consumer surplus at R is.

$\int_0^R B(dR/dt)$(d(dR/dt)). Planning board must maximize the present discounted value at an extraction rate $R_t$(t > = 0) in order to maximize the integral:

$$\int_0^t e^{-rt}\{S_0^{R_t}\text{B (dR/dt) d(dR/dt)}\}\text{ dt,} \tag{11}$$

subject to the constraints $S_t = S_0 - \int_0^t R_\tau d\tau$ and $R_t$ and $S_t$ both greater than and equal to 0 for t > = 0.

A feasible extraction policy (rate of extraction being positive at all times) is efficient if and only if $S_0 = \int_0^\infty R_t dt$. We obtain the optimal solution as (d$p_t$/dt)/$p_t$ = r. As desired, this efficient optimal extraction policy satisfies the condition of $S_0 = \int_0^\infty R_t dt$.

### 2.4. Reserve exhaustion at infinite time horizon

Assuming demand is linearly increasing with time, so $p_t$ = A– B.$R_t$, where A and B both > 0. As long as reserves are positive, market equilibrium price holds and spot price will be given by.

$$p_t = p_0 e^{rt}. \tag{12}$$

Then, we obtain $R_t$ = A/B – [($p_0 e^{rt}$)/B], which gives $R_t$ = 0 for $p_t$> = A.

Reserve will be exhausted at future time. Time of exhaustion, T, will be at A where demand falls to zero.

http://dx.doi.org/10.5772/intechopen.88125

Initial price $p_0$* is given by $p_0{}^*e^{rt}$ = A, and then, exhaustion time, T, is obtained from the relation:

$$\int_0^T TR_t dt = \int_0^T \{(A/B)-(p_0*e^{rt}/B)\}dt = S_0. \tag{13}$$

The above analyses show that mineral resources (including lean-grade ores) are exhaustible.

### 2.5. Independent versus dependent sample data

Natural phenomena are very complex and may not follow simpler mathematical and probabilistic assumptions necessary for analyses, estimation, tests, and decisions needed by scientists, technologists, and managers. For example, rocks and ores in the crust are heterogeneous, anisotropic, and inelastic materials at different scales such as microscopic, mesoscopic (hand specimens), and megascopic (outcrops) levels which make the sample size (REV) vary according to the scale of heterogeneity.

Geological processes are often nonlinear which needs complex linearization pre-transformations for mathematical and statistical analyses. Data are often non-Gaussian, need Gaussian/normal prior transform for simpler univariate (scalars) or multivariate (vectors) statistical analyses, are temporally/spatially dependent, and need complex time series analyses (wavelet and/or geostatistical analyses). Because of these complexities and of space constraints, these advanced methods are omitted here, but we give a few summary tables for guiding the readers.

More details on the nonlinear, stationary, nonstationary, and seasonal time series at time-, frequency-(Fourier), and time-frequency/wavelet domains are given in Sahu [15]. Whereas nonlinear models are in use for several hundreds of years, time series models are in use for about 100 years and were popularized by Box and Jenkins in 1970 and 1984, but wavelet models are currently very popular [25] (**Tables 1** and **2**).

| Statistics | Fourier transform | Wavelet transform |
|---|---|---|
| Variance | Frequency spectrum | Wavelet spectrum |
| VAR ($E(X^2)$) | S(X) (w) = X X* = $(\|X\|)^2$ | W(X)(s, τ) = $(\|X\|)^2$ |
| Covariance | Cross-spectrum | Wavelet cross-spectrum |
| COV (E(XY)) | S(X,Y) (ω) = XY* | W(X,Y) (s,τ) = XY* |
| Correlation coefficient | Coherency | Wavelet coherency |
| r = E(XY)/{E(X*2) E(Y*2)} | γ = S(XY)(ω)/Sq.rt.S(X)(ω)S(Y)(ω) | $\Gamma = W(XY)(s,\tau)/\sqrt{W(X(s,\tau)\,W(Y)(s,\tau)}$ |
| Coeff. of determination | Coherence | Wavelet coherence |
| $r^2$; r as above | $\gamma^2$; γ defined as above | $\Gamma^2$; Γ as defined above |

**Table 1.** Equivalent statistics for statistical (independent samples), Fourier (time series) (dependent samples), and wavelet analyses (dependent samples).

| Stationary parameters | Tests | Linear prediction | Linear methods | Limitations |
|---|---|---|---|---|
| **Univariate statistical: random IID scalars as inputs** | | | | |
| **Mean, variance, correlation coefficient** | t, Chi-sq, F, r, R^2 LR | Conf. limits of the parameters | ANOVA, ANCOVA | Samples, independent |
| **Multivariate statistical: random IID vectors as inputs** | | | | |
| **Mean, cov−/corrln. Matrices** | T^2, Chi-sq, F, R^2 LR, partial corrln. | Conf. limits of the parameters | PCA, FA, CCA, MANOVA LDF, CLASSIF, MANCOVA | Samples, independent |
| **Univariate time series models with random IID scalars as inputs having constant lag data** | | | | |
| **Mean vector, acv, acf, spectrum** | WN, AR, MA, ARMA nonst: ARIMA | Conf. limits of the parameters | ARMA(p,q), d = 0: ARIMA(p,d > 0,q) | Other information lost |
| **Multivariate time series models with random IID vectors as inputs having constant lag data** | | | | |
| **Mean vector, acv, acf, spectra** | WN, AR, MA, ARMA Tr. Fn. forecasts, upgrade | One- and multi-step times | Tr. Fns with delay | Heterogeneity |
| **Univariate geostatistical models with random IID scalars as inputs having local stationarity (d = 0)** | | | | |
| **Mean, sill, range, nugget** | Model validation | Nil | Interpolation, kriging | Extreme values |
| **Multivariate geostatistical models with random IID vectors as inputs having local stationarity (d = 0)** | | | | |
| **Multiple spectra, coherency** | Model validation | Nil | Interpolation, kriging, co-kriging | Extreme values |

**Table 2.** Comparison of linear statistical, time series, and geostatistical models for RVs.

## 3. Mineral exploration

Finite natural resources including metallic and nonmetallic ores, hydrocarbons, and industrially usable earth materials have been exploited since the emergence of mankind on the earth, and hence, surface and near-surface deposits are facing exhaustion/near exhaustion. The ever-increasing demands for improving living standards have resulted in present rapid rate of exploitation of ores and hydrocarbons with corresponding critical deterioration of the environment inducing high health risks. Therefore, at present there is an acute need for intensifying exploration for new concealed and/or partially concealed ore deposits at greater depths.

Geochemical exploration is one of the cheapest and important tools for detecting such hidden targets both at regional and local scales in the initial stage and at local scale during the mine development and production stages. Optimal exploration of any region must include integration of relevant available information on geology, geochemistry, and geophysics in order to delineate homogeneous regions within heterogeneous crustal rocks, ores and mineralizations, and surfaces. Homogeneity can be achieved through clustering by nearest neighbor (NN), fractal, or inverse distance weighting methods.

At the exploration stage when sample data are sparse, the purpose is to delineate, either directly by the element or through its pathfinder(s), regional and local (positive) anomalies which generally include many ore targets, whereas at the mine development and/or production stages, large sets of closely spaced data become available for better statistical analyses using time series, fractals, etc. to estimate regional or local background(s) for delineation of such targets. Future actions at exploration stage include recommendations of intensive exploration and a few test-drilling at high positive anomalies to make mine feasibility studies, if desired, and to suggest mining and ore beneficiation methodologies to be adopted. In production stage, the purpose of geochemical exploration would be to find additional local targets for drilling and mining (if it is ore grade). Thus geochemical exploration plays very important roles in identifying subsurface targets of ore as well as hydrocarbon deposits and their subsequent exploitation.

### 3.1. Univariate statistical methods

Exploitable ore deposits and hydrocarbon resources are associated with crustal rocks and rock systems at shallow depths. Valuable mineralizations are emplaced along with the formation of these rocks (syngenetic deposits) or later formed/intruded into these rocks (epigenetic deposits) through their pores and/or fractures (faults). Some deposits may be deformed, modified, or weathered by several earth processes and at several later times (tectonic phases). For mathematical and statistical purposes, ore deposits can be characterized for geochemical purposes and subsequent statistical analyses. Rocks comprise $p > 1$ number of mineral phases which sometimes include valuable ore minerals and waste/gangue minerals. Minerals can be geochemically characterized by their elemental, molecular, and isotopic constituents that form major (>10%), minor (1–10%), or trace (<1%) amounts in them on volume or weight basis.

However, the main problems for statistical analysis of geochemical data are as follows:

i. Rank is not full rank (i.e., not p; but p-1 or less) as total sum on volume/weight basis is 1.0/100% (closure constraint) which introduces spurious negative correlations among the constituents both among and within samples. The data are not independent for usual statistical studies but are dependent [10, 11]. Besides, samples with 0 and 100% should be deleted as they are not admissible for further analyses [18].

ii. Frequency distribution of constituents is on weight/volume basis and not on number basis as needed for statistical analysis. The distributions are not Gaussian but binomial ($\log(x/(1-x))$-normal) for major and minor components and Poisson having lognormal distribution for trace components [18].

Unfortunately, binomial and Poisson distributions, being discrete, are not that easily amenable for statistical analyses and hypotheses tests as it is for Gaussian distributions. For large sample ($N > 50$), binomial and Poisson distributions can be Gaussianized using the central limit theorem, but this method is sampling intensive and hence very costly to be practical and useful. A better alternative as suggested by Sahu [10, 11, 15, 16, 18] is to use log-odds pre-transform as Gaussian distribution of (fractional) major or minor constituents which reduces to simple log transformation of fractional values for trace constituents. These log-odds transforms also simultaneously make all the constituents independent of others in the rock/ore sample and thus eliminates the rank problem/closure constraint.

So, all univariate normal/Gaussian (UNIV) (if N > 20) and multivariate statistical analysis (MND) with full-rank models can be applied (if N > 33) as desired. There exist other pre-transformations in literature such as log ratio [7, 20], but this is still not full rank and uncorrelated as desired in statistical theory [21]. After Gaussian pre-transformation of geochemical constituents in the sample using log-odds, linear or nonlinear statistical methods of time series or spatial series or time-spatial series (which include geostatistics (= ARIMA(p,1,q)) as a special case) or fractal/multi-fractal which are nonlinear models can be used to estimate regional/local background values from which regional/local anomalies can be identified [26].

Further actions should include:

i. Intensive geochemical investigations (including a few drillings) made at positive anomalies (prospects)

ii. Feasibility for exploitation

iii. Financial decisions for optimal mining and beneficiation (of low-grade ores) and mine remediation measure [14, 17].

### *3.1.1 Anomaly detection using log-odds, i.e., log $(x/(1-x))$; $0 < x < 1$, where x is fractional assay of geochemical constituent by linear and nonlinear modeling*

| Exploration and initial development stages | Advanced development and production |
|---|---|
| **A. Graphical univariate multivariate methods** | |
| Background: < av. x + 2σ | Less than av. vector x + 2σ |
| Anomaly: not less than av. x + 2σ | Not less than av. vector x + 2σ |
| Breakpoint in cdf of X in probit | Breakpoint in cdf of vector X (probits) |
| Above 3 for pathfinder element(s) | Above 3 for pathfinder elements |
| **B. Fractal/multifractals (nonlinear methods which have their first term as linear)** | |
| c-N, c-L, c-A, and c-V use log–log plots | c-N, c-L, c-A, and c-V use log–log plots |
| Breakpoints give regional and/or local background values from which local and/or regional anomalies and targets can be identified for further exploration | |
| Interpolation: by inverse distance, spline | Interpolation: by inverse distance, time series signal, fractals/multifractals, RBF, SV, kriging, spline |
| Singularity index gives robust backgrounds that are stable which yield good targets | |
| Fractal analysis in conjunction with neuro-fuzzy-genetic analysis and other soft computing techniques can yield very good to excellent ore targets | |

Elemental concentrations in multicomponent (p > 1) rock and/or ore mineralizations are in different proportions (major >10%; minor 1–10%; trace levels <1%) which either have binomial distributions for major/minor proportions or reduce to Poisson distribution for trace constituents. These concentrations are spatially/temporally, as well as within each sample, dependent, and hence statistical methods dealing with independent data are not strictly applicable. Independence of various constituents within any sample can be achieved through log (odds ratio) transformation for each constituent which simultaneously achieves Gaussian probability distribution, so that univariate normal (UNIV) theory and multivariate normal distribution (MND) theory (using linear theory) become applicable. Multivariate ($p > 1$ or vectors) theory reduces to univariate theory when a number of random variables become one ($p = 1$, or scalar), so univariate model is included under multivariate model. Population parameters like mean vector ($\mu$) and covariance matrix (D) are then estimated for each population, and appropriate statistical tests are performed to take suitable statistical and geological decisions. Ore mineralization exists in the crust and is a three-dimensional (3D) static (not time-varying) phenomenon which can be modeled as a multivariate normal system through Gaussianization of log (odds) of constituent fractions present in rock/ore/soil samples. Mineralization processes are extremely complex, nonhomogeneous, and nonlinear to be analyzed directly. Therefore, these must be partitioned into homogeneous subsystems having strong dependence within but very weak interactions among the populations (groups). Two types of univariate hypotheses include (i) sample mean that belongs to a given population (null or $H_o$) which is tested against alternative hypotheses ($H_1$) and (ii) variances that are homogeneous (equal, $H_o$) or not ($H_1$). Two kinds of errors made for any decision are (i) Type I or $\alpha$ error of rejection of $H_o$ when true and (ii) Type II or error of acceptance of $H_1$ when false. It is prudent to test more powerful ($1-\beta$) error being maximized for a given Type 1 error; these procedures are extendable for MND models.

Multiple and polynomial regressions (correlations) are strictly univariate model having only one random error but generally included under multivariate model because matrix methods are essential to solve them as it is necessary in MND analyses. However, F tests cannot be reliable for polynomial regressions, since powers of variable cannot be Gaussian when the variable is Gaussian. However, any power function relation of dependent variable(Y) with predictor variable(s) (X) can be linearized using log(Y) pre-transformation for statistical analyses.

### 3.2. Multivariate statistical analysis

This is appropriate for analyzing multiple correlated measurements (random vectors) made on one or more samples and on one or more (homogeneous) populations/groups. If a p-variate ($p \times 1$) Gaussian random vector (X) is measured on N-independent samples of a population, then the mathematical model has multivariate normal distribution (MND) which is characterized as all linear compounds of the variables being also MND as the rank of vector X may/may not be equal to its order (p). MND methods are classified on the basis of the number of populations (one or more) and number of sets of random vectors (one or more).

Four classes thus form:

- One population and one set of variables: MND methods are principal component analysis (PCA), factor analysis (FA), and cluster analysis (CA).
- One population but more than one set of variables: MND methods are multiple regressions (correlations), polynomial regressions, and canonical correlation.

- One set of variables but more than one population: MND methods include MANOVA, discriminant functions (DF: linear, quadratic), and classification function (CF).
- More than one set of variables and populations: MANCOVA.

Constituents of rocks/ores/soils are constrained to 1.0 or 100% which form a mathematically induced dependence structure (not geologically interpretable or meaningful) but have a binomial/Poisson distribution with heteroskedastic (or unequal) variances. Gaussianization of marginal distributions are often performed, but this does not guarantee that the joint distribution of random vector belongs to MND. So, MND theory must be tested through which all linear combinations (esp. principal components, multiple/partial/canonical regression components) of the random measurements are MND. However, MND theory is very robust, and if N is fairly large, then the random vector may be accepted to have MND.

The parameters of a multivariate r.v. can be estimated for any homogeneous population by its mean vector (μ) and dispersion (correlation) matrix (D/R) using MLE. Null hypothesis of homogeneity of population mean vectors (H2) conditional on homogeneity of dispersion matrices is given by T^2 test:

$$T^2 = N\,(m - \mu)^T\, D - 1\,(m - \mu), \text{ with } (N - p)T2/(N - 1) \tag{14}$$

is F, p,(N-p) distributed with m being sample mean. Homogeneity test for covariance matrices (H1) of different populations is more involved (Box test).

Matrix operations are essential for multivariate analysis. Matrix A is a rectangular array of numbers with p rows and q columns, where a (i,j) is its ijth element. Addition and scalar multiplication are straightforward, but matrix multiplication requires that the number of columns of the pre-matrix must be equal to the number of rows of the post-matrix; otherwise multiplication is not defined.

If AB = C exists, then the elements c(i,j) = sum over r (a(i,r) x b(r,j)).

But in general multiplication is not commutative and AB not equal to BA, so pre- or post-multiplication of the matrix must be specified. However, multiplication is associative: A (BC) = (AB)C = ABC. Transposed matrix A^T has the rows and columns interchanged in A, so (AB)^T = B^T A^T.

If X and Y are two conformable column vectors, then their inner product is given by X*Y.

If A is (m × n) matrix, then A.x is a column vector. A* is conjugate transpose of complex matrix A, then (AB)* = B* A*, and so on.

Rank of matrix is the number of independent columns (or rows) in the matrix. A square matrix of order m is nonsingular, if its rank (m) is less than its order (p). A unique inverse matrix A-1 exists if A is nonsingular, then AA^-1 = A^-1 A = I (identity matrix). If unique inverses of A and B exist, then (AB)^-1 = B^-1 A^-1; also (A)^-1 = AT and (A*)^-1 = (A^-1)*. Unitary matrix, A* has A*.A = I = A.A* and so, A* = A^-1. If A is a real nonsingular, square matrix is said to be orthogonal if A^T A = A A ^T = I, so A^-1 = A^T. A Hermitian matrix has A^-1 = A*. Elementary operations on columns (rows) of a matrix can give simpler form to interpret and compute, but its rank is preserved.

Determinant of a square matrix A (=$a_{ij}$) can be obtained by expanding the element ($a_{(i,j)}$) of a row (column) by multiplying its cofactor and summing over all elements of the row (column) or by multiplying the eigenvalues of A. Generalized inverse of a singular matrix (rank less than its order) is denoted as A-, and then $A A^{-} A = A A^{-}$ is not necessarily unique.

$A^{-}A = H$ with $H^2 = H$ (idempotent).

If $A^{-} = A$, we get det A = r(A) = r(H) = trace(H). If $A^{-}$ exists, then $r(A^{-}) \geq r(A)$.

Quadratic form (Q) of matrix plays an important role in MND analysis and is given by $Q = X^T A X$ with $A = [(a_{(ij)} + a_{(ji)})/2]$ which is a symmetric matrix. If ($X^T A X$) is >0, it is positive definite (pd), =0 (null), and it is negative definite (nd) and semi-definite if 0 is included in the product. For any nonsingular linear transformation, Q remains definite and invariant. Every positive definite matrix A can be decomposed into $CC^T$ where $C^T$ is inverse(C^-1) of the linear transform matrix. A necessary and sufficient condition for A to be positive definite (p.d.) is that its determinant is positive. This summary on multivariate analysis is based on Sahu [16].

#### *3.2.1. Principal component and factor analysis*

This method is for single population and one set of random variables. Original vectors in p-dimensional space are linearly transformed to a smaller m-dimensional subspace of principal components which are orthogonal. Mathematically, a real symmetric covariance (correlation) matrix is diagonalized (all correlations become zero) such that the principal diagonal yields the eigenvalues (variances) along the orthogonal eigenvectors (directions). In factor analysis, some of the smaller nonsignificant eigenvalues are deleted as negligible error components without losing information. The retained eigenvectors are rotated orthogonally in the lower common factor space (m < <p), so the new correlations (loadings) become easily interpretable (either near 1 highly loaded/correlated or near zero loadings if uncorrelated) as rotated factors. Rotation of orthogonal factors in the lower space is made by standard varimax program [27]. Cluster analysis can be made to obtain homogeneous groups by using similarity or distance matrices, but this process is rather empirical and needs great care for accuracy.

Eigen-structure of correlation (dispersion) matrix (R or D) is achieved through powering the matrix to a very high index (say 64 or 128) so that the largest eigenvalue dominates over the rest and corresponding eigenvector is obtained by a few iteration. The effect of the first eigenvalue is subtracted from the matrix to obtain the residual matrix which is again powered to high index to get next eigenvalue and eigenvector. This sequence is continued till all information of R (or D) are extracted and residual matrix becomes zero matrix. However, before running the eigen-structure analysis, null hypothesis R = I must be tested for statistical significance by a chi-square test with p(p-1)/2 d.f. at 0.05 level.

$$\text{The test quantity is given by} - \left[(N-1) - (1/6)\,(2p+5)\right] \ln\,(|R|). \qquad (15)$$

Spectral decomposition of matrix gives.

$$R = \Lambda 1 \ v1\ (v1T) + \Lambda 2\ v2(v2T) + \ldots + \Lambda\ p\ vp(vpT) = \Sigma\ Rj \text{ over all } pj. \quad (16)$$

If m components are found to be statistically significant, then the rest (p-m) components are noise and are deleted. So, total variance explained is sum R(j) of first m components, and rank of R is now m (<< p).

Multiplication of all eigenvalues gives |R| and sum of all eigenvalues is called trace of R.

Principal factors (f(j); j = 1 to m) are computed dividing the retained eigenvectors by the square root of their eigenvalues. Thus each factor becomes equally important as the other with a variance of 1 for all j. Factor structure S = V (Λ)^ -1/2 and predicted R by all factors is S*S^T; residual error is R- S*S^T. The number of significant principal components (m) retained as factors is the most important.

A chi-square test of determinant of residual matrix, res (A) with (p-m)(p-m-1)/2 d.f., is given by.

$$-\{(N-1)–1/6(2p+5) - 2/3\ (m)\}\ \ln\left[|R|/\{\Pi \text{ of m eigenvalues } (p - \text{sum m eigenvalues}))/ (p-m)\}(p-m)\right]], \quad (17)$$

which is tested at 0.05 level.

Another method is to plot jth eigenvalue vs. j to get inflection point giving m factors or to plot standard deviation of cum. Eigenvalues are computed on independent replicate samples of size N from the same population vs. j to get a minimum at which cum. Eigenvalue of 85% or more gives m. This second procedure, given in 1973 by the author, is a second-order criteria for deciding the common factor space (m) [16]. Varimax rotation is absolutely necessary to eliminate non-interpretable intermediate loadings in the range of 0.2–0.5 in any unrotated eigenvector of principal component.

Factor j is interpreted by the rotated loadings in the jth rotated eigenvector as follows:

**i.** Absolute value of loadings close to unity is statistically significant and identifies the factor in terms of the input variables, and loadings near zero are nonsignificant and do not contribute to this factor (but may identify some other factor on which they are strongly loaded).

Correlation matrix can be computed over N samples to give R-mode R showing correlations among the random variables or over the p variables to give Q-mode R showing correlations among N samples. However, either R or Q correlation matrices have the same information and hence give finally the same inferences/decisions. But the order of R in R-mode is p << N; hence computationally R-mode analyses are preferred/cheaper. The rotated eigenvalues are different from the variances from corresponding eigenvalues, although the total variance (= Cumulative Eigenvalue) of m(<< p) retained factors is conserved by orthogonal rotations as can be easily demonstrated by matrix theory [16].

### 3.2.2. *Multiple regression (correlation) and canonical correlation*

Multiple (including polynomial) regression (correlation) yields linear prediction of dependent (criterion) variable(Y) from the knowledge of the predictors (X). The slope of regression line b = (Var x)^ -1 Cov (x,y) if X and Y are scalars (univariate analysis), which is extended to vector random variables as b = Cov(x)^ -1 Cov (x, y) if X is a vector random variable and Y is a scalar random variable. Multiple correlation exists if multiple correlation coefficient R is statistically significant, and R2 indicates the sum of squares explained by predictors and (1-R2) indicates noise sum of squares.

F test can be made with (p-1) and (N-p) as degrees of freedom. However, since elements of X are mutually correlated (not independent), the effect of each element of X on Y is highly confounded and not possible to correctly interpret. Partial correlations remove the effects for other elements mathematically to give correct inferences for correlation of Y with ith element, x (i), of X, and hence, it is preferred over multiple correlations. In canonical correlation, two or more sets of variables are needed: one set is criterion, the other set predictor, and the third set control which can be kept mathematically constant. In contrast to principal component analyses, the eigen-structure is computed along the maximum covariances (not along maximum variances). The total correlation matrix R (with y as the pth r.v.) is partitioned into X of order (p-1), and hence we get the real nonsymmetric matrix as R22-1 R21 R11-1 R21 which is the product of two real symmetric matrices: B = R22 and A = R21 R11-1 R12.

Mathematically we solve the eigen-structure of (A – ΛB) = 0 or of eigen-structure of B-1 A V = V (Λ). Eigen-structure of B-1 A can be done through two stages:

**i.** Eigen-structure of real symmetric matrix B to give Λ1 and obtain B^-(1/2).

**ii.** Eigen-structure of symmetric matrix (B^-1/2 A B^-1/2) gives Λ2 and eigenvector U2. Eigenvalues of B^-1A are the same as that of (B^-1/2 A B^-1/2), and hence sought vector, V, is given by B^-1/2 U2.

Statistical significance of diagonal elements of canonical eigenvalues (Λ2) can be assessed as follows:

**i.** Proportion explained by Λj = Λj/ (trace Λ2)

**ii.** Bartlett Lamda statistic = Product (j = 1 to p2) of (1 – Λj), where p2 is dimension of predictor vector

The null hypothesis that criterion and predictor sets are uncorrelated is assessed through chi-square with p(1) x p(2) d.f. as: - [(N-1) – (1/2) (p(1) + p(2) + 1)] ln (Λ).

If null hypothesis of no correlation is rejected, then the effects of the first canonical root (Λ1) is subtracted; the rest p(2) -1 canonical roots tested as:

Product (j = r + 1 to p(2)) (1 – Λj) as a chi-square with degrees of freedom, d.f. = (p(1) -1). (p(2)-1).

$$\text{Chi} - \text{square} = -\{(N-1) + \tfrac{1}{2}(p(1) + p(2) + 1)\} \ln\ (\Lambda 1\ \text{residual}). \tag{18}$$

This test is continued until nonsignificance is achieved.

**i.** A thumb rule: a canonical correlation < absolute 0.30 is statistically nonsignificant and hence dropped.

Multiple regression with standardized variables z can be written as z (hat) = b1 z1 + b2 z2 + ... + bp zp and multiple correlation coefficients Rp. 1,2,..., (p-1) = R are similar to product–moment correlation coefficient (r) having the range of [−1,+1] for linear regression of scalars, but R has a range from 0 to 1.

R2 explains a major part of the variance of criterion and (1-R2) gives the error variance of regression. Therefore, F test with (p-1) df in numerator and (n-p) df as the denominator is applicable for the quantity R2 (N-p)/(1-R2)(p-1).The (p*p) correlation matrix R can be partitioned into R11 with order (p-1), and the last criterion (scalar) z(p) has a variance of 1.0. The multiple slope vector b = R11^-1 R12.

However, high values and high significance of any bj do not imply true importance of zj since other predictor z's confound the multiple correlation slopes. Hence partial correlation of criterion with a zj keeping all other predictors mathematically constant is absolutely necessary for any statistical/geological inference.

Polynomial regression is similar to multiple regression, but powers of predictor of random variables and the interaction terms are included. High degree of polynomial regressions is very difficult to interpret, and also if X is Gaussian, then its powers and interactions cannot be Gaussian, precluding the use of F test for the regression equations. So, unless theory dictates such polynomial regression, it should be avoided, and in any case, the degree should be as low as possible (say, second order).

Multiple partial correlation matrix R2.1 = {r21.jk} = residual r22.jk/(res r22.jj x r22.kk) 1/2. (19)

For a trivariate-random variable, system res r22 = 1- r212 and res r23.j = res r23.j/(1 – r212)1/2. So, r21.3j = res r23j/(1 – r212)1/2, a well-known result in statistical theory. The output of partial correlation analysis can be arranged as: R = [(R21/ R32.1) (R21.3/R33)].

An example of partial correlation would clarify many of these concepts developed above.

The following random variables were measured in 33 thin sections from 33 sandstone samples.

- The variables were X1 = phi long axis of grains (which has Gaussian distribution).
- X2 = matrix percent.
- X3 = porosity percent as reported by Griffiths in 1967, (p.468).

The multiple correlation matrix R was found with r12 = .8813**, r13 = −.7094**, and r23 = −.66771**. Here, ** means statistical significance at 0.01 level.

We compute partial correlation r13.2 = (r13 – r12 x r23)/(1– r122)1/2. Partial correlations between X1 X3, X1 X2, and X21.3 are similarly computed, and we get r21.3 = .6439**, r31.2 = −.3862**, but r23.1 = −.2222 NS instead of −.6671**. [Here, superscript ** means statistical significance at 0.01 level.]

Therefore, r23.1 is nonsignificant indicating X2 and X3 are truly independent (uncorrelated) rather than correlated or possessing a negative multiple correlation. This fallacy of multiple correlation coefficients must always be noted and true inference must be sought through computation of partial correlations.

**Comments**: Although X1 has a Gaussian distribution, X2 and X3 possess closure constraints (ranging from 0 to 100% or 0 to 1 as fractions) and not Gaussian but binomial. X2 and X3 should be Gaussianized by the prior transformation log (xj %/ (100% − xj,%)) for j = 2,3. Multiple correlations should have been computed with original X1 and the new transformed-Gaussianized X2 and X3 variables.

### *3.2.3. MANOVA: discrimination and classification*

MANOVA is similar to ANOVA for vector random variable X. In ANOVA (scalar r.v), two types of tests are necessary to test equality of main effects:

**i.** When interactions are nonsignificant, the interaction variances are pooled with error variance, and a pooled error variance is calculated to yield the F test.

**ii.** When interaction variance is significant, then its variance is used to test main effects by F test.

In MANOVA, treatment variance is divided by the pooled error variance to give F test since interaction variance is nonsignificant. But if interaction variance is significant, then MANCOVA methods are used to test main effects (F test) by dividing treatment variance by interaction variance (not error variance).

Populations (groups) are not necessarily homogeneous in mean vectors and covariance matrices. Two situations can arise.

**a.** Covariance matrices are homogeneous, and testing is done to find homogeneity of mean vectors (H2 test; linear discriminant functions (LDFs) and MDFs as hyperplanes) or otherwise.

**b.** At least one covariance matrix is different; we have to use nonlinear quadratic hypersurfaces (QDF) to delineate regions of each population. If both the mean vectors and covariance matrices are utilized together, then the procedure is called classification.

We decompose an ith vector of kth group $X_{(ki)}$ from grand mean m as $x_{(ki)} = X_{(ki)} - m = (m_{(k)} - m) + (X_{(ki)} - m_{(k)})$, where $m_{(k)}$ and m are the mean vectors for kth population and all populations, respectively. So any data is the sum of main effects (among-group) and within-group deviations $(X_{(ki)} - m_{(ki)})$.

The SSCP is then $\Sigma\, x_{(ki)}\, x_{(ki)}^T = \text{Sum}\,(m_{(k)} - m)\,(m_{(k)} - m)^T + \text{Sum}\,(X_{(ki)} - m_{(k)})^T$, summed over i = 1 to $N_{(k)}$ and k = 1 to g groups. Symbolically, T = A + W, where only two matrices are independent because of closure constraint. We get $W^{-1}T = W^{-1}A + I$, having only one independent matrix, $W^{-1}$ A, for further analysis as identity matrix (I) is a constant. If covariances (correlations) among the groups are equal ($H_1$ true), the dispersion among the groups is (D (A)) = A/(g-1), and dispersion within groups is D(W) = W/(N-g) where N is the total data over g populations.

The null hypothesis $H_2$: $\mu_{(k)} = \mu$ for all k = 1,…,g. and m = (sum $X_{ki}$ over all I and k)/N. Rao [28] proposed F test as follows:

$$s = \left\{\left(p^2(g-1)^2 - 4\right)/\left(p^2 + (g-1)^2 - 5\right)\right\}^{1/2}; n_{(1)} = p(g-1); n_{(2)} \\ = s\left[(N-1) - (p*(s-1)+1)/2\right] - (p(s-1)-2)/2. \tag{20}$$

Let y = $(|W|/|T|)^{-s}$. Then, $F(n_{(1)}, n_{(2)})$ = ((1-y)/y) $(n_{(2)}/n_{(1)})$ and tested for statistical significance.

$H_2$ true, if F test is nonsignificant, means all mean vectors are equal.

*3.2.4. Linear discriminant function*

For two groups, g-1 = 1; hence there can be only one LDF, linear discriminant function (hyperplane). But for multi-groups g-1 is more than one, so we can have several LDFs, some of which may not be significant (should be dropped), but we also need the angles between the accepted (significant) LDFs (hyperplanes).

The retained LDFs form a subspace within the original p-dimensional space, and samples may be projected onto this subspace for visual studies. Optimal solution is to maximize the ratio W-1A (nonsymmetric real matrix) in the common discriminant subspace defined by vector v s.t.; the ratio of (Λ) = (vT A v/ VTWV) is maximized with the constraint vTv.

The maximum values are the eigenvalues of W^-1 A: that is we solve (W^-1A) V = VΛ.

Since W is full rank, W^-1 is unique and can be decomposed as U Λ1 U^T, so W-1/2 = U(Λ1^-1 U^T).

Then, eigenvalues of W^-1 A = eigenvalues of W^-1/2 A W^-1/2, but B has a different eigenvector U2.

Since B is symmetric, its eigen-structure is U2Λ2U2^T, and the eigenvector matrix V of W-1A is obtained as V = W^-1/2 U2 and has eigenvalue matrix Λ2.

The number of LDFs to be retained are obtained by statistical significance tests for elements Λ2,j where j = 1 to (g-1) or p whichever is minimum (= rank of W-1A matrix). The importance of jth discriminant function (if retained as significant) can also be judged by the ratio of Λ2,j/trace Λ2 where this ratio ranges from 0 to 100%.

Also, each Λ2,j can be tested as a canonical correlation of discriminant vector vj with any population (group) as the criterion (Y). The eigenvectors in V should be normalized (to vector with magnitude 1), and the angle between the ith and jth linear discriminants J(i) and J(k) is given by (θ(i,j)) = Cos-1(v(ik). v(jk)). These angles are not necessarily orthogonal since W-1A is a nonsymmetric matrix. Discriminant scores which can be computed as v(i)T x(jk) for each retained eigenvector ji and xjk are the kth sample of jth group.

These scores can be projected onto the common discriminant subspace for visual perusal. A chi-square test of significance of discrimination amount for remaining m-k discriminants after accepting the first k significant discriminants can be assessed and tested as $-(N-(p + g)/2)\ln \Lambda^*$ with df = (p-k) (g-k-1), and $\Lambda^*$ is the product of $1/(1 + \Lambda j)$ for j = (k + 1 to m).

This chi-square test should be nonsignificant to stop analysis. Usually two discriminants are most useful for visual representation of projection of LDFs as straight lines in the discriminant space, but 3D projections can be made if three discriminants are significant and required [16, 29].

### 3.2.5. *Quadratic discriminant function (QDF)*

If at least one covariance matrix is unequal among the groups, then pooling of covariance matrices to give a common (homogeneous) covariance matrix is inadmissible. Then, discriminant is nonlinear and hypersurface given by $\mu_{(1)} D_{(1)}^{-1} \mu_{(1)}^T - \mu_{(2)} D_{(2)}^{-1} \mu_{(2)}^T$ which reduces to LDF if $D_{(1)} = D_{(2)} = D$ and QDF = LDF = $(\mu_{(1)} - \mu_{(2)}) D^{-1} (\mu_{(1)} - \mu_2)^T$ as was derived under LDF theory. If a number of samples $N_{(1)}$ and $N_{(2)}$ are large, LDF is sufficiently robust for applications. Also, for QDF,F, test is inapplicable to find its significance.

MANCOVA methods (not discussed here) are more involved but necessary and proven to be useful for multi-element ores and for multiple populations (groups) in order to discriminate and/or classify.

## 4. Some applications

Statistics in the technology of twenty-first century and along with current capability of computers will be essential, beneficial, and most useful to mining and mineral processing industries. A log $(x/(1 - x))$ pre-transformation of fractional concentrations ($x$; $0 < x < 1$) yields the desired independence and Gaussian pdf of each constituent in the rock or ore. This is necessary for characterization/estimation of parameters of each pdf and for hypotheses of tests and inference. Univariate and multivariate statistical models are used for single and multiple pre-transformed random variables, respectively. These models are useful for geochemical exploration, mining, mine planning, mineral processing, and beneficiation and for marketing such that maximal profits with minimum environmental damage can be achieved to obtain sustainable economic and societal growth.

Some applications of statistical (Gaussian) technology to mineral industry are listed below (not exhaustive) for getting the feel of different scenarios involved:

**Exploration**

**a.** Detection of positive anomalies for further intensive search

**b.** Detection of negative anomalies for use as sinks for toxic materials/elements

c. Decision on pathfinders using factor model for single-element and canonical correlation for multiple-elemental ores

d. Decision on feasibility of mining using open-pit or underground mining method

**Mining**

a. Development of optimal mine plans using concentration contours and risk analysis

b. Decision on lateral and vertical extensions to present mine plans

**Beneficiation**

a. Decision on lower limit of assay for grinding

b. Decision on optimal beneficiation process and system to be installed

c. Locating high-grade ore zones for conservation for later blending and marketing

d. Waste management decisions and related operation planning

**Marketing**

a. Optimal classification of ore grades by separation and/or beneficiation/blending

b. Marketing of nonmarketable ores in situ or in dumps by blending and/or beneficiation in the future

**Example 1:** Estimation of ore reserves and average assays.

In the past spatial distributions of assays and their pdfs were not accounted for, and simple calculations yielded these quantities based on geometry of ore body and arithmetic averaging of assays within mineable ores. The geology of syngenetic ore deposits produces uni- or multi-metal binomial/Poisson distribution having homogeneous variances, whereas epigenetic ore deposits are likely to possess bimodal (low assays in host rock and higher assays in ore zones) distributions that need separate treatments. A log $(x/(1 - x))$ pre-transformation of fractional assays achieves linearity, Gaussianity, and homoscedasticity of variance with elimination of spurious negative correlations among the constituents [15, 16].

### 4.1. Syngenetic deposits

The lower limit of assay value for mineable ore is given by lower than 95% confidence limit of mean which should be marketable as well. Multi-metal ores are converted to single metal ore through addition of equivalent prices of these metals or by using a principal component of the mineable/marketable elements. The associated risk factor should be evaluated for use of lower than 95% confidence limit.

### 4.2. Epigenetic deposits

Bimodal Gaussian pdfs which can result as the mean in host rocks may be much less than in ore body. Univariate/multivariate discriminant function (LDF) easily separates these two modes for separate calculation of reserves and average assays. Since mineable ore zones may lie within ore body, or partly in ore body and host rock, the geometry of mineable ore zone can be complicated. A 3D mine model is necessary to delimit mineable zones, and this may be achieved through a computer system. The associated risk factor for mining should also be computed as indicated above.

### 4.3. Spatially correlated samples on log(x/(1 − x)) basis

This situation often arises in development and production stages when a large number of geochemical data becomes available. The data is having signal (mineralized assay) and noise (random errors) and time (spatial) series model which separated the signal needed for average assay computation from the Gaussian noise giving the confidence limits to the average assay.

Models may be nonstationary (ARIMA (p,d,q) which is made stationary (ARMA(p,0,q) by differencing the data 'd' times. Integration of location data over the spatial domain gives the total volume and of signal gives the average assay. Such integration can be done block-wise to obtain block reserves and block average assays. More details of general time series modeling are given in Sahu [15]. Geo-statistical models are a special case of time series models belonging to the ARIMA (p,1,q) if assay values are linearized prior to analysis but otherwise are generally nonlinear since log (x/(1 − x)) pre-transformation of fractional concentrations was not performed.

**Example 2:** Identification of pathfinders

Pathfinders include minerals, molecules, elements, and isotopes and are very useful for exploration of uni- and multi-metal ores. These are characteristically easy to recognize, have higher concentrations than sought element and higher occurrence frequency, and can be analyzed cheaply and easily at much less time. Correlation matrix, R, of pre-transformed fractional constituent data is then computed using R-mode analysis (cheaper and faster than Q-mode). The correlation coefficients could be strongly +ve ($r > .50$), strongly −ve ($r < -.50$), and weak ($-.50 < r < +.50$).

Factor analysis of R (without weak correlations) provides rotated factors that are statistically significant to yield pathfinder(s) loaded strongly on high positive and/or negative correlations of constituents. Pathfinders can also be identified through the use of partial correlations, but this method is computationally more intensive, and decision may be confounded (nonunique). Multi-element or multi-mineral ores would require canonical correlation analysis with the mineable metals/minerals taken as criterion vector and other sets as predictor and/or control random vectors. FA in such cases may not be useful unless criterion vector has only one random constituent which is mineable. More details on finding geochemical pathfinders can be obtained from Sahu [16].

**Example 3:** Mine feasibility.

Sustainable mining operations must insure that the expected profits/year remains positive and substantial for meeting cash flow and other financial commitments. Mining operations with profits depends on high sale value of high-grade and beneficiated low-grade ores to marketable grades. Mining costs include mine operation, transportation, beneficiation, and disposal of mine wastes with remediation.

Main geological factors are:

**i.** Total reserve (W tons)

**ii.** Proportion of high-grade, W(H)

**iii.** Proportion of low-grade W(L)

**iv.** Proportion of gangue W(G) = 1- W(H)-W(L)

**v.** Assay for high-grade ore A(H)

Of these five, three factors are independent but (iv) factor W(G) is not independent of W(H) and W(L). A(H) and A(L), assays of low-grade ores, are two additional independent factors (total of five factors).

There are ten economic factors including profit per year, P, which must be positive; life of mine (L = W/ PR (production rate per year in tons)); sale price per ton of marketed ore S(H); capital cost of mine operations, C(M); capital cost of beneficiation plant, C(B); rate of interest, r; efficiency of beneficiation, e; per ton running costs of mine, R(M); beneficiation, R(B); and waste disposal R(D).

Economic analysis for mine operations having beneficiation processing results in profits = P:

$$P = [W\ (W(H) + W(L))\ S(H)/L)–(C(M) + C(B)\ (1 + r)/L)–(W\ R(M)/L)–(W - W(H).A(H) \\ - W(L)\ e\ (A(H)–(A(L)/L)\ R(D)–(W\ R(B)/L) \tag{21}$$

Above equation is a complex nonlinear one which cannot be linearized. If W(L) is negligible (zero), then mine-site beneficiation would be unviable and associated costs C(B) and R(B) would become zeros. Smaller mines with less low-grade ores would need pooling of nearby small mines for establishing a combined beneficiation plant to be operated jointly by these mines with proportional cost and profit sharing per year [14].

**Example 4:** 3D modeling and mine planning.

Fast, efficient, and up-to-date computer systems having links to end users are essential for this purpose.

The following items seem to be useful:

1. Fast transmission of databases, maps, and sections to central processor online and/or offline

2. Preparation of 3D maps using GIS technology and generation of desired sections for planning and mine operations. Offline transmission of this to all subcenters

3. Optimal plans for mine transport, blending and beneficiation, and timeframe of works

4. Marketing plans, expected profits vs. actual profits, doubling of assets in <5/6 years
5. Planning for new biddable targets and for extensions to present mines
6. Monitoring environmental damage and plan mitigation of such damages
7. Optimal computer architecture on faster transfer ratio in several separate computer systems acting in parallel to yield distinct outputs from a single input to the whole system
8. Mine closure plan, settlement of personnel, equipment disposal, mitigation of ground- and surface water damages, and corporate social responsibility (CSR) for the locality and country

**Example 5**: Mine sustainability.

Environmental hazards and associated mitigation costs are site-specific and hence, will depend on local geology, topography, and climate. Mineralogy and geochemistry would affect emissions on metallurgy, pollution by toxic metals, and leakages from dumps/tailings. Environmental degradation is due to poor production efficiency and poor innovations.

We can achieve sustainability in mining industry through the following:

1. Market incentives with pollution prevention, focusing on management system (MS)
2. Mine closure plans with Environmental Impact Assessment (EIA) and SIA (S = social) at all stages
3. Bonds which should be issued to clean up pollution after mine closure
4. Obtaining environmental and social performance indicators, risk assessment for environmental management (EM), and use of life cycle assessment (LCA), for technology choice and EMS

## 5. R&D efforts for resource augmentation

R&D efforts for exploration of new deposits, invention of natural and/or technological substitutes, and waste treatment/disposal are omitted. Since high-grade ores are comparatively rare in occurrence and will be exhausted in finite terminal time, T, in-house R&D efforts in mining industry should primarily concentrate on optimal utilization of associated lean-grade ores that are not directly marketable as well as on treatment/disposal of associated waste products. There are two ways to upgrade lean-grade ores and market the upgraded products through in-house R&D efforts as given below:

- Optimal marketing of lean ores by blending with appropriate amount of (locally available or imported) high-grade ores having assay value of a1
- Optimal marketing of lean ores by technological beneficiation with or without blending with high-grade ores having assay value of a1

### 5.1. Ore blending

We assume for convenience that the exploited ore body has only one type of marketable grade and only one type of nonmarketable lean-grade ores with associated waste materials of little market value. The analysis can be easily extended to two or more types of marketable ores and lean ores.

Let price per ton of ore be p and the assays and weight fractions of high-grade and lean-grade ores be a1, f1 and a2, f2, respectively, so that waste fraction is (1- f1 – f2). We have to crush the high-grade and lean-grade ores to suitable optimal size for blending so that the mix becomes homogeneous and can yield a stable grade, a* > a(m), the minimum marketable grade necessary for marketing purposes.

Then the average assay of blended mixture is.

$$a* = (W1\ a1 + W2\ a2)/(W1 + W2), \tag{22}$$

where W1 and W2 are, respectively, the weights (in tons) of high-grade and lean-grade ores in the mixture. The blended ore can be sold at a market price p per ton. The total sell value of blended ore would be (W1 + W2)p.

The blended assay, a1*, must be greater than the minimum marketable assay, a(m), and then the triangle law of proportional lengths of sides provides the minimum W1 value as (a(m)–a2)/a1. However to be safe, a1* should be made 5% more than required minimum, a(m), value so that blended ore is not rejected and we can use a minimum W1 value of (1.05 a(m) – a2)/a1). If this minimum amount of W1 of the high-grade ore is not available, then we have to optimize the mixing weights W1 and W2 for maximizing the profits by using constrained Lagrange multiplier or may have to import the deficient quantity of high-grade ore.

### 5.2. Ore beneficiation

Beneficiation of lean ores can be performed using physical, chemical, and biochemical methods, and the optimal technological parameters for providing maximum present value to marketed products should be found by several experiments and using appropriate response-surface experimental designs. The lean ores are to be finely grounded using jaw crushers, ball or rod mills, to optimal grinding size to maximally liberate the ore minerals for beneficiation [17].

The optimal mix of beneficiation products can be estimated by using Lagrange multipliers to maximize the net present value (defined as sell price minus cost price) of beneficiated products. Details of applicable optimal beneficiation schemes and the associated plant designs are specific to the mine, ore type, and ore characteristics and, hence, cannot be discussed as a general system theory. Therefore, the different ore beneficiation procedures are not discussed here.

### 5.3. Cutoff grade estimation

With continued rapid rate of utilization of minerals and metals for societal growth and explosive rise in world population, it has become imperative to mine minerals with lower grade and at greater depths which induces increased costs of extraction and processing to make these marketable. Hence, there is a greater need for conservation of these invaluable nonrenewable finite mineral resources as well as for the preservation of fragile ecology and environment. Balancing these two opposing concepts of maximal utilization of ores and sustainable societal growth is the critical need of the hour [1–3, 5, 14, 25].

Mineral resources are characterized by their unique geological setting and genesis, as well as their spatial distribution which greatly influence the optimal extraction of these nonrenewable resources. Mining industry becomes sustainable with consistent long-term profit accruals over the life span of the mine.

This dictates that extracted ores can be marketed with reasonable profit, with sale price (s/ton) exceeding the cost of production (c/ton). The cost of production includes many factors such as mining, blending, beneficiation, transport of ores and wastes for their marketing, and safe disposals, respectively. Sale price (s/ton) of marketable ore is highly unpredictable due to volatile demand and supply of ores, government policy, technological innovations, substitute products, etc.

The cutoff assay, x(C), is defined as the fractional assay (x) of resource above which the extracted product is marketable (x(M)) and above the break-even assay (x(B)) defining the equality of sale (s/ton) and cost (c/ton) prices of the produce. Unfortunately, break-even assay, x(B), is not very useful as a cutoff grade since at this mining strategy profits become nil and mine becomes unsustainable. However, x(B) does provide the upper bound to the cutoff grade x(C) for ore extraction. Profits accrue if the extractable grade is reasonably above the marketable grade x(M) with the sale price (s/ton) greater than the production cost(c/ton) and with an extraction rate that maintains long-term sustainability of mine (at least till the end of mine life or ore exhaustion).

This strategy would induce a cutoff grade, x(C), much lower than the x(B) assay value but should be equal to the minimum assay value, x min, or near zero assay value or waste materials. The lower-grade materials with assays less than the assigned cutoff grade, x(C), are not mined and left in situ as un-mined blocks and pillars.

The optimal cutoff assay, x(C,O), therefore, must lie satisfying the following sequence:

$0 < x \min < x\ (C,O) < x\ (B,R) < x(B)$ or $x(M) < x$ maximum $<1.0$. Under the static model, even though.

x(C,O) variation has a high range of assays (between x min and x (B)), it can be optimally estimated using two factors of pdf of $\lg (x/(1-x))$ or $\lg(x)$ which is Gaussian and the ratio of sales to cost prices (s/c).

Under a dynamic model, however, these parameters are time-varying and hence have to be estimated for each time period of mining extractions, and hence, the procedure of estimation of x(C,O) becomes more complex and time-consuming. Sale price is much more volatile than the cost of production as it depends on supply and demand position, vagaries of technological innovations, market substitutions, government interventions, and management policies. Dynamic x(CO) must account for these dynamic changes for optimizing the current profits and the future expected profits. All profits must be brought to a comparable level using the standard techniques of reduction using net profit value (NPV) [3, 14] to the present state of time origin. Characterization of pdf of fractional assays in the ore body can be made by measuring assays of a large number (N > 50) of independent REV [30] samples/cells/blocks collected in the 3D space over which the resource exists.

From these sample data using the standard statistical methods, the arithmetic mean, median, variance, standard deviation, etc. can be easily obtained/computed. However, it is well known that the fractional assay pdf is lognormal [6, 9, 15, 16], and hence all statistical parameters and hypothesis tests must be made on the transformed Gaussian random variable, log(x), where log stands for common logarithm of assay value, x.

Computing on log(x) basis, we obtain the mean (u) which is median of log(x), and standard deviation of log(x) ($\sigma$) for cutoff, x(C), and optimal cutoff, x(C,O) estimations.

In the case of high-valued ores like diamonds, U, REE, Au, Ag, etc., the optimal cutoff is to x(B). In static analysis the geologic, assay distributional and economic factors are assumed to be constants over time, while in dynamic analyses, these parameters are time-varying and have to be estimated after each time unit of ore extraction (say, quarterly, half-yearly, or yearly as felt necessary). Dynamic modeling, although more involved, would yield much greater profits than the simpler static analysis.

### 5.4. Relation between break-even, x(B), and cutoff, x(C), assays

The main goal in mining is to maximize the profits by sale of mined and/or beneficiated ores [5]. Break-even assay, x(B), is very important in mining industry as it delimits the profitable ores from nonprofitable mineral resources.

This concept largely depends on the ratio of sales (s/ton) and cost (c/ton) prices of ore extraction of the marketable ore grade, x(M). Thus, using the pdf of fractional assays, x, if x(C) is zero, or x min, we obtain, for cdf F(x) of assay values, x as

$$F(x(B))\, s = c \tag{23}$$

$$\text{or } F(x(B)) = 1-(c/s) \tag{24}$$

At some positive cutoff grade x(C) > 0 and/or x min, we similarly get

$$(1-F(x(B))\, s = (1 - F(x(C))\, c, \tag{25}$$

or

$$F(x(B)) = 1-(1-F(x(C))\, (c/s). \tag{26}$$

Eq. (2) can be reversed to obtain F(x(C)), as a function of F(x(B)), as

http://dx.doi.org/10.5772/intechopen.88125

$$F(x(C)) = 1–(1–F(x(B)) (s/c). \tag{27}$$

Break-even grade, x(B), obtained by $F^{-1}x(B) = N^{-1}x(B)$ and, similarly, optimal cutoff grade, x(CO), can be obtained by inverting F(x(CO)) or cutoff grade x(C) by $F^{-1}x(C) = N^{-1}x(C)$.

In dynamic models forecasted values of (s/c) ratios are needed which is achieved by linearizing Eq. 3 and adding random error terms to sale (e(s)/ton) and cost (e(c)/ton) prices, as

$$\log\,(1–F(x(C)) = \log(1–F(x(B)) + \log(s/c) + \log(e(s)) - \log(e(c)). \tag{28}$$

The predicted values of log (1–F(x(C)) = Y(hat) from time series model equation (Eq. (25)) can, then, be inverted to obtain x(C) as {1 – Exp (Y(hat))}. But in this paper, dynamic models (for forecasting s and c values to get Y (hat)) are not investigated in details and not pursued further. From Eq. (24), it is obvious that F(x(B)) is the upper limit to cdf of cutoff grade F(x(C)), and if (s/c) is 1.0, then F(x(C)) = F(x(B)).

We can obtain the lower bound to cdf of cutoff grade F(x(C)) as the minimum assay in the resource, x min, since fractional assay always lies above x min for any lognormal pdf.

Thus, F(x(C)) lies in the range from (x min) to (x(B)) in the following sequence:

0 < x min < x(C)/ x(C,O) < x(B,R) < x(B) < x(M) < x max<1.0; (x(M) being the marketable grade and x(B,R) the break-even grade for mining with risk factor at alpha confidence level). The optimal cutoff grade, x(C,O), will lie in the range from (0/x min) to (x(B))/(x(B,R) or x(M)); and x(C,O) strongly depends on the (s/c) ratio as well as the lower (x min or 0) and upper, x(B,R) or (x (B)) bounds of x(C).

**Table 3** indicates some typical values of lower and upper bounds of x(C) for various (mineable/ economic) sale-to-cost price ratios (s/c) for a lognormal assay pdf, x, (i.e., log(x) is Gaussian).

Geologists and mining engineers have been using cutoff grade empirically, determined by their experience or by simple thumb rule and without using the concept of pdf of assay distribution being lognormal in the ores.

| (s/c) ratios | 1.15 | 10 | 100 | 1000 | 10,000 | Remark |
|---|---|---|---|---|---|---|
| x min or zero | 0.3 | 0.01 | 0.001 | 0.0001 | 0.00001 | Lower bound |
| F(x(B)) | 0.84 | 0.99 | 0.999 | 0.9999 | 0.99999 | Upper bound |

Standardized random variable, z, can be calculated as z = [log(x) – μ]/σ; hence we obtain x min = Exp [μ – σ. z min] and x max = Exp (μ + σ.z max). The values, F(x(B)) and x min/zero grade, form the upper and lower bounds of cutoff grades, respectively. However, the upper bounds to F(x(C)) given in **Table 4** would be more useful, in practice, rather than those listed in **Table 3**.

**Table 3.** Feasible break-even grade, x(B), and x min or zero grade < x(C)/ x(C,O) assume lognormal (log-Gaussian) pdf for log(x) with mean, μ, and standard deviation, σ.

| x | 1 | 2 | 3 | 4 | 5 | 6 | 10 | Remarks |
|---|---|---|---|---|---|---|---|---|
| F(x(B)) | 1 | 0.5 | 0.33 | 0.25 | 0.2 | 0.167 | 0.100 | Upper bound |
| F(x(C)) | 0 | 0 | 0 | 0 | 0 | 0 | 0 | Lower bound |

**Table 4.** Values of upper bounds, F(x(B)), and lower bounds, x (min) or zero, to the cutoff grade, F(x(C)), for different economically feasible (s/c) ratios, greater than 1.0, as obtained by Eq. (2).

## 5.5. Profit maximization

Assuming sufficient proportion of marketable ores, (1 –F(x(B)), with sale price (s/ton) greater than the production cost (c/ton), exists for mining until exhaustion (mine life), economic viability is obtained if.

$$W.(1\text{–}F(x(B))\ s > w.c. \tag{29}$$

The optimal cutoff grade, x(C,O), should be searched for in the range of profitable cutoff grade range (i.e., between x min and x(B)), which can be equivalent to z (C,O) value within the range of z(x min) to z (x(B)) calculated on the basis of lognormal mean (μ) and standard deviation (σ) of the log(x) pdf. Then the total profit, P, can be calculated as.

$$P = W.(1 - z\ (x(C)).(s - c). \tag{30}$$

The optimal cutoff grade, x(C, O), would be less than x(B) and can be obtained through risk analysis (sale price of material in the range of x(B) to x(C,O) is equal to the cost of mining up to this lower optimal grade, x(C,O)) [5, 16]. The resulting new break-even grade, x(B,R), taking risk at alpha level, would be less than the original break-even grade, x(B), without taking any risk, that is, x(B,R) < x(B). Using cumulative unit normal, N(0,1), statistical tabular values, we then obtain [5, 16].

$$[N(z(x(B)) - N(z(x(C,O))]/[N(z(x(C,O)) - z(x\ \text{min})\ \text{or}\ z\ (-\infty)] = (c/s), \tag{31}$$

which can be easily computed at the mining office for static or dynamic modeling to obtain x (C,O) value, as required by the mine management. Compare Eq. (31) with Eq. (27) for calculation of F(x(C) assuming mining of ores at zero assay value as the cutoff grade.

Dynamic model calculations must be updated from time to time after a fixed time period of mining operations. In addition, forecasting techniques can be used to predict future sale price (s/ton) and future cost price (c/ton) using time series modeling parameters estimated from corresponding past time series data [15] to obtain c/s or s/c ratios as needed.

These forecasting techniques are complex and model parameter dependent, and hence not pursued further here. In the case of large mines, blending and/or ore beneficiation processes are usually employed to upgrade lean ores to marketable grades for profits as well as for waste utilization to protect ecology/environment.

http://dx.doi.org/10.5772/intechopen.88125

Dynamic optimization of profits is mine and ore specific and cannot yield general optimal cutoff grade for any specific ore type (say, iron ore, copper ore, gold ore, etc.). Hence, dynamic optimization must be done periodically in every large mine and for every block as needed. In addition, mining company must adhere to the National Mineral Policy and Mining Laws to have safe and uninterrupted operations.

## 6. Conclusions

i. Fractional constituents of rocks/ores form multicomponent system having less than full rank to be statistically meaningful for analyses and geological inferences. A $\log(x/(1-x))$ pre-transformation of x data is shown to eliminate rank problem and Gaussianizes/normalizes (independent (uncorrelated) of sample data) the random variables/vectors for linear statistical inferences, for geologically appropriate decisions, using univariate and multivariate statistical model procedures. More complex analyses such as dependence (correlated) samples in temporal, spatial, and spatiotemporal domains are not used here for simplicity and for lack of space.

ii. Geochemical concentrations ($x$; $0 < x < 1$) are rarely point data but are averaged over 1, 2, or 3D space and attached to the center of the sample. Also, measurement base is rarely numbers, as needed for statistical analyses, but the averaged sample values over length, area, or volume (weight) are used as random variables located at the center of sample volume (REV). Fractional constituents are spuriously as well as complexly negatively correlated because of closure constraint for the total measure (1.0 or 100%) and possess nonconstant variance over their mean value because these are binomial (multinomial) for major and minor levels which reduces to a Poisson distribution for trace components.

iii. Spatial patterns of these constituents are usually heterogeneous as mineralization intensities vary over local and regional scales. Homogeneity can be achieved through sampling at scales of representative elementary volume (or greater volume) and over local scales through NN clustering techniques.

iv. Anomaly detection is a very complex task as it depends on accurate determination of local or regional threshold levels. The associated risks for exploration of different types of anomalies as well as for specific individual targets must be evaluated.

v. Mineralization involves highly complex nonlinear geological processes, so simple univariate statistical (linear) approaches with graphical methods are neither unique nor optimal. Multivariate approaches are also linear but are much better (however, becomes unrealistic if the sample data are spatially/temporally correlated). Nonlinear methods such as nearest neighbor, fuzzy logic (FL), genetic algorithm (GA), and soft computing (SC) techniques are more complex to use, not dealt here, but can be useful.

vi. During exploration stage useful multivariate methods could be multiple and partial correlation, factor (principal component) analyses, canonical correlation for identifying pathfinders and delineating anomaly zones, two-group linear discriminant functions, and multi-group linear discriminant functions for identifying ore from wastes and for delineating ore zones.

But during the development and production stages, it is prudent to identify different categories of ore and zones such as

- Marketable high grades
- Blended marketable grades

| (s/c) | 1.0 | 1.1 | 1.2 | 1.25 | 1.3 | 1.4 | 1.5 | 1.6 | 1.7 | 1.8 | 1.9 | 2.0 |
|---|---|---|---|---|---|---|---|---|---|---|---|---|
| **F(x(B))** | F(x(C) | | | | | | | | | | | |
| **.005** | .005 | | | | | | | | | | | |
| **.010** | .010 | | | | | | | | | | | |
| **.015** | .015 | | | | | | | | | | | |
| **.020** | .020 | | | | | | | | | | | |
| **.025** | .025 | | | | | | | | | | | |
| **.030** | .030 | | | | | | | | | | | |
| **.035** | .035 | | | | | | | | | | | |
| **.040** | .040 | | | | | | | | | | | |
| **.045** | .045 | .000 | | | | | | | | | | |
| **.050** | .050 | .000 | 0 | | | | | | | | | |
| **.100** | .100 | .010 | 0 | 0 | | | | | | | | |
| **.150** | .150 | .070 | 0 | 0 | 0 | | | | | | | |
| **.200** | .200 | .120 | .040 | 0 | 0 | 0 | | | | | | |
| **.250** | .250 | .175 | .100 | .060 | .025 | 0 | 0 | | | | | |
| **.300** | .300 | .230 | .160 | .120 | .090 | .020 | 0 | 0 | | | | |
| **.350** | .350 | .285 | .220 | .190 | .145 | .070 | .025 | 0 | 0 | 0 | | |
| **.400** | .400 | .340 | .280 | .250 | .220 | .160 | .010 | .040 | 0 | 0 | 0 | 0 |
| **.450** | .450 | .395 | .340 | .310 | .285 | .230 | .175 | .120 | .065 | .010 | 0 | 0 |
| **.500** | .500 | .450 | .400 | .370 | .350 | .300 | .250 | .200 | .150 | .100 | .050 | 0 |
| **.600** | .600 | .340 | .280 | .250 | .220 | .160 | .100 | .040 | x | x | x | x |
| **.700** | .700 | .230 | .160 | .125 | .090 | .020 | x | x | x | x | x | x |
| **.800** | .800 | .120 | .040 | x | x | x | x | x | x | x | x | x |
| **.900** | .900 | .010 | x | x | x | x | x | x | x | x | x | x |
| **1.00** | 1.00 | x | x | x | x | x | x | x | x | x | x | x |

Beyond (s/c) ratio of 2.0, the F(x(C)) values will be essentially, either, x (minimum) or zero; hence these values have not been given in this table. The optimal cutoff grade (as given by Eq.(V.5,-2), F(x(CO))) lies between F(x(min)/zero) and F(x(B)) values.
Values of x(C) = F^-1(x(C)) = N^-1(lg(x(C)); similarly, we obtain x(CO) = F^-1 (x(CO)) = N^-1(lg(x(CO)).

**Table 5.** Cumulative value of fractional cutoff grade (assay), F(x(C)), as calculated by Eq. (27), for different ratios of sales/cost prices, (s/c) ratios, and for different cumulative values of fractional break-even grade, F(x(B)), as calculated by Eq. (26), is given in **Table 5**.

- Beneficiated marketable grades
- Low-grade ores forming future resource
- Waste material

To achieve these delineations, we use multi-group linear discriminant function approach.

Then, optimal mining, blending, beneficiation, and marketing operations would maximize profits, social aspirations, and ecological/environmental protection/remediation for greater sustainability.

**vii.** Optimal utilization of unmarketable lean-grade ores helps to accumulate additional capital for economic and social growths and reduction in costs of waste treatment and disposal, thereby improving the health of local inhabitants, as well as helps in the conservation of high-grade ores for better sustainability and for utilization in the future.

**viii.** Estimation of cutoff grade, x(C), or optimal cutoff grade, x(CO), involves many complex geological, spatial assay distributional, pdf of assays in rocks/ores, and economic factors such as sell price (s/ton) and production to marketing costs for ores plus disposal costs of wastes (c/ton).

Estimation is simpler for static models where these random variables (rvs) are Gaussianized/normalized using pre-transformation, log(x), and are time-invariant constants but highly involved when these parameters vary with time and have to be updated at short intervals (dynamic models). Estimation of cutoff grade can be obtained by solving the nonlinear equation for static models (see **Table 5**) and inverting the obtained F(x(c) or F(x(CO)) using the standard unit-normal cumulative statistical tables giving N(0,1), i.e., N^ -1(0,1).

Dynamic models do not have a global solution for cutoff grade since it has to be updated at every time interval of mining operation but yield much greater overall profits to the mining industry.

## Author details

Basanta K. Sahu

Address all correspondence to: bas.sahu37@gmail.com

Department of Earth Sciences, IIT Bombay, Mumbai, India

## References

[1] Sahu BK. Geology applied to sustainable development of mineral resources (Key-paper). In: International Seminar, Souvenir Volume; ISM Dhanbad; 2013. pp. 77-83

[2] Sahu BK. Mining and sustainable development of non-renewable mineral resources. In: Proceedings of the Seminar on Sustainable Development on Mineral Resources and Earth Processes, New Delhi; 2014. pp. 4-7

[3] Sahu BK. Optimal mining and marketing of lean grade ores. SGAT Bulletin. 2015;**16**(1):1-6

[4] Sahu BK, editor. Proceedings of the Geomodelling Workshop; IIT, Bombay; 1983. 274p

[5] Sahu BK. Optimal cutoff grades (assays) in mining for profit optimisation. In: International Conference on Mineral Processing Technology (MPT, 2016), Pune; 2016 (submitted). 7p

[6] Sahu BK. Probability density function (pdfs) of minerals and/or chemical constituents in rocks and ores for sustainable growth. In: MPT 2013 International Seminar, Bhubaneswar; 2013 (in press)

[7] Aitchison J. The Statistical Analysis of Compositional Data. London: Chapman & Hall; 1986

[8] Sahu BK. Multivariate statistical methods for mineral exploration using geochemical data. In: Vistas in Geology. Bhubaneswar: Utkal University; 2013. pp. 7-16

[9] Ahrens LH. The lognormal distribution of elements. Geochimica et Cosmochimica Acta. 1954;**5**:41-73; 121-132

[10] Sahu BK. Mineral deposit modeling. Mineralium Deposita. 1982;**17**:99-107

[11] Sahu BK. Statistical inference for geochemical and petrographic data. In: Proc. Rec. Res. NW India; Vadodara, Gujarat: M.S. Univ.; 1995. pp. 59-64

[12] Dasgupta PS, Heal GM. Economics and Extraction of Exhaustible Resources. Cambridge, UK: Cambridge University Press; 1980. 501p

[13] Pindyck RS. Optimal exploration and production of nonrenewable resources. Journal of Political Economy. 1978;**86**:841-861

[14] Sahu BK. Geology and economics for mining low-grade ores. SGAT Bulletin, Bhubaneswar. 2012;**13**:72-74

[15] Sahu BK. Time Series Modeling in Earth Sciences. Lisse, Netherlands: Balkema; 2003. 284p

[16] Sahu BK. Statistical Models in Earth Sciences. Hyderabad: B.S. Publications; 2005. 211p

[17] Sahu BK. Optimal liberation of ore minerals for ore beneficiation. In: International Seminar on Mineral Process Technology (MPT2010); Jamshedpur; 2010. pp. 3-8

[18] Sahu BK. Evaluation of iron ore prospect in BIF setting. In: Proc. Intl. Sem. Iron Ores, SGAT, Bhubaneswar; 2008. pp. 2-11

[19] Sahu BK. Statistical methods for geochemical exploration of ore deposits. In: Vistas in Geology. Bhubaneswar: Utkal University; 2013. pp. 1-6

[20] Egozcue JJ et al. Isometric log-ratio transformation for compositional data. Mathematical Geology. 2003;**35**:279-303

[21] Carranza EJM. Analysis and modeling of geochemical anomalies using log-ratio transforms. Journal of Geochemical Exploration (JGE). 2011;**110**:167-185

[22] Le Cam L. Asymptotic Methods in Statistical Decision Theory. New York: Springer; 1986. 742p

[23] Le Cam L, Yang L. Asymptotics in Statistics. New York: Springer; 1990. 180p

[24] Hotelling H. The economics of exhaustible resources. Journal of Political Economy. 1931; **39**:137-175

[25] Sahu BK. Prediction of permeability in clastic reservoirs and aquifers using stochastic and wavelet theories for sustainable fluid production. In: Intl. Conf. Adv. Water Resources; CAS Panjab Univ; 2013. 17p

[26] Delarer ST. Delineation of mineralization using c-V fractal method. Journal of Geochemical Exploration (JGE). 2012;**118**:98-118

[27] Kaiser HF. The Varimax criterion for analytic rotation in factor analysis. Psychometrika. 1958;**23**:187-200

[28] Rao CR. Linear Statistical Inference and Applications. New York: Wiley; 1973. 625p

[29] Sahu BK. Multi-group discrimination of river, beach, and dune sands using sphericity statistics. Mathematical Geology. 1982;**14**:577-586

[30] Bear J. Dynamics of Fluid Flow Through Porous Media. New York: Elsevier; 1972. 372p